Deborah Lupton, Clare Southerton, Marianne Clark and Ash Watson
The Face Mask In COVID Times

Deborah Lupton, Clare Southerton,
Marianne Clark and Ash Watson

The Face Mask In COVID Times

A Sociomaterial Analysis

DE GRUYTER

ISBN 978-3-11-111665-5
e-ISBN (PDF) 978-3-11-072371-7
e-ISBN (EPUB) 978-3-11-072379-3

Library of Congress Control Number: 2021931902

Bibliographic information published by the Deutsche Nationalbibliothek
The Deutsche Nationalbibliothek lists this publication in the Deutsche Nationalbibliografie;
detailed bibliographic data are available on the Internet at http://dnb.dnb.de.

www.degruyter.com

Contents

List of Illustrations

Preface

Deborah

2020 for me has been a year of different kinds of face masks, for different purposes. The first time I ever saw mass mask wearing was when I visited Japan in January 2020 for a family holiday. The COVID-19 crisis was just on the horizon but had not yet affected Japan to any great extent. In the streets of Tokyo, most people were wearing masks in public: not due to fear of novel coronavirus contagion, but because it was winter, and this was an accepted custom to prevent against the spread of seasonal colds and influenza. I felt ill at ease when walking around in Tokyo with my face uncovered. Were Japanese people looking at me and my family and judging us foreigners because we were not behaving in socially acceptable ways by shunning mask wearing? Each time I coughed or sneezed, I wondered if the Japanese people near me were subtly edging away from me, feeling disgust at my poor manners and lack of hygiene. It was impossible to tell. The masks on their faces made it difficult for me to assess their feelings.

We returned home to Canberra, the national capital of Australia, a city that for weeks on end was filled with choking bushfire smoke from the horrendous 'Black Summer' of 2019–2020, affecting many parts of Australia for months on end. I could see, smell and taste the risk of the smoke-polluted air. We purchased an air filtering unit for our home and kept all our windows closed despite the stifling summer heat. I ordered a special face mask from the UK to filter out the tiny particulates from the fires near our city that were blown in by strong winds and settled in the basin bordered by mountain ranges in which Canberra lies. The mask was ugly and chunky. I tried to stay home and out of the thick, dark air that some days ranked among the worst in the world for dangerous air quality levels. I wore the mask for brief periods when I went outside to see how my garden plants were faring in the long, very dry summer. I noticed some other people walking their dogs along my street, also wearing masks. This was the first time I had ever seen people wear face masks of any kind in my neighbourhood. I hated the airless sensation of the thick mask on my face, and how hot and uncomfortable it felt. But I knew that breathing in this smoke could have ill effects on my lungs, so I needed to wear the mask.

By March 2020, cases of COVID were beginning to increase rapidly in Australia, and restrictions and lockdowns were implemented nationally. At first, we were advised not to wear face masks. Later in the pandemic, health authorities changed their minds. Face masks became part of the complex precautions in which Australians were expected (or in some cases, required) to engage to 'stop the spread' and 'get out of lockdown'. I heard that disposable face masks were

https://doi.org/10.1515/9783110723717-001

Figure 1: Hand-made fabric masks ordered online. Photo credit: Deborah Lupton.

becoming hard to find and went online to order some reusable cloth masks. I found some pretty hand-made vintage floral-patterned masks on Etsy and spent a small fortune on them for myself and my daughters (Figure 1). I now have a drawer full of face masks. But they remain unworn. Thus far, our city has suppressed the spread of the novel coronavirus very well, and there is no real need or official requirement to wear a mask. I have seen very few people out in public with masks on. I tried putting on one or two of my fancy new masks at home to see how they felt. I still disliked the breathless feeling they give me (although they do look pretty). I hope I never have to wear them regularly.

Ash

As I line up for the grocery store check-out in my eastern Sydney suburb, my half-weekly basket of fresh goods hanging from my arm, I am struck by how many kinds of face masks surround me. There are a few sky-blue medical masks, the same as the one I am wearing, which I bought in bulk after stumbling upon one last box in an office supplies store. Many more masks are not as intentionally disposable. Most, in fact, seem unique and quite beautiful creations: cloth masks in block colours or floral print, some gently padded across the nose, with matched elastic that loops behind the ears or ribbon-like legs that tie around the back of the head. These masks are soft, personal, deliberate; there seems to have been different thought put into their making than the throwaway one I had selected. Mine, I suppose, came together through a factory-sized system of automated machinery rather than the steady hands of someone who sat in their living room, thinking of my face.

The models on my Instagram ads match, literally now, from head to toe. I contemplate buying a T-shirt with a mask-wearing emoji emblazoned across the chest, the cotton a coral-pink and images in fluorescent yellow. I imagine it as a bizarre memento I could wear ironically over the summer, showing it off to my family over Facetime, and then keep in a shoebox in the cupboard for a later decade. I flinch before clicking it into my shopping cart. I hesitate to tempt fate, to assume that there will be a later time, that I make it through unscathed, that my disparate family remain healthy, that I might enjoy remembering this period through a T-shirt.

Clare

There's usually a mask tucked in my handbag; one sits in the cupholder of the car. I can often find another one sitting next to the bowl where I keep the car keys. Occasionally, when I put on a raincoat or a jacket I haven't worn for a while, I'll find a mask in the pocket. I know this isn't the right way to store them. I read the articles about mask 'hygiene' routines, and yet they still end up stuffed everywhere, haphazardly, usually the result of being flung off as soon as possible.

'How many masks should I have' I type into Google one day. The answer is varied: maybe five? *InStyle* magazine tells me to treat my masks like underwear, a confusing mental image. But, like underwear, I do have favourites. 'I brought the good masks,' I say to my girlfriend as we get to the supermarket and she's visibly relieved. 'Good. I hate the other ones.'

The 'good' ones feel pretty flimsy, and I often wonder if they do much at all, but they're the most breathable ones we have. They're made of a thin blue material that's very light and the most notable sensation while wearing it is the heat within the mask, rather than the struggle to breathe. I see fewer people wearing them now around my local area of Sydney. There was a time, it seemed brief, that they were everywhere in the supermarket, on the faces of the shoppers around me. Now I feel a bit silly sometimes when I wear mine. If I catch the eyes of someone passing me, I wonder if they think I'm sick or paranoid.

Earlier in the year, I spent several days at a Sydney hospital visiting a sick family member. Mask-wearing in the hospital was mandatory, along with temperature checks at the door and strictly timed visits. While I found myself confronted by the daily reality of mask-wearing, which Australians living outside the state of Victoria have avoided mainly, it was somewhat of a relief to no longer feel the ambiguity around masks that persists in many other places in Sydney. I often would assess each situation anew ('Is this a place where I should mask-up? How crowded will it be? Will other people be wearing masks? Will I feel uncomfortable or embarrassed if I wear one?'). In the hospital, the mask felt expected and ordinary, even as I felt relief when I could remove it.

Marianne

In late March 2020, after the pandemic had been declared, a care package arrived to my Sydney home from my Mum, who lives in Canada along with the rest of my family. It contained three hand-sewn masks, one each for my partner, myself and our two-year-old daughter. The 'mini-mask' for my daughter was made from colourful fabric featuring her favourite characters, delighting her until she actually put it on. It is now repurposed, along with ours when we aren't using them, for her dolls and her make-believe games (Figure 2). The face mask now a part of my little daughter's imagining of the world.

I remember being struck by this act of care, this form of touch, performed over such great distance. More masks followed in varying fabrics and designs, in case we found them more comfortable, in case they better suited the climate. Care took on different textures. Yet my experience of *wearing* a face mask is … complicated. It's so profoundly physical *and* deeply social. Masks feel hot, no

Figure 2: Child's face masks used for make-believe play. Photo credit: Marianne Clark.

matter how lightweight, and they make everything sweaty. I'm always self-conscious about the moisture on my face when I remove my mask after stepping off the bus or exiting a shop. But I also feel somewhat comforted by the presence of the mask on my face. I liken it to the feeling of wearing a wetsuit in the ocean, ever so slightly fortified against the raucousness of the 'out there'.

At the same time, masks carry such social weight. I'm sure I've been assessed as paranoid in situations where I'm the only person wearing one. But I have also been quietly ashamed after stepping onto a bus and realising I don't have my mask in my pocket.

Seeing my Dad and Step-Mum wearing a face mask for the first time during a Zoom call while they were out in a public place in their Canadian town, for some reason startled and saddened me. Reflecting later, I came to realise the sight of them in masks reminded me of their – and my own – vulnerability, and once again of the distance between us. The sight was also a jarring reminder about how much the world, and our lives, have changed. My mother-in-law's wistful lament, during a conversation about masks, still echoes. 'I miss seeing smiles,' she said.

While my understandings of masks are shaped by these specific embodied and social experiences, so too are they shaped by living in and through this pandemic, by the endless numbers signalling rising cases and deaths worldwide. By the statistics telling devastating stories about the inequities of our world. Through this, I've come to wonder what else might we do? Surely much needs to be done. Masks cannot protect us all from all of these complexities, but perhaps as a symbol of care, they are a good place to start.

1 Introduction: The Shifting Meanings and Practices of Face Masks

Introduction

When US President Donald Trump announced via a tweet in the early hours of 2 October 2020 that he and his wife Melania had tested positive for COVID-19, news reports and social media outlets were inundated with commentary. Many points were made about Trump's position on face mask wearing as a preventive measure against infection with the novel coronavirus (SARS-CoV-2). For months, Trump had notoriously downplayed the dangers of COVID to his fellow Americans and had discounted the need to wear masks. For him, mask wearing was a sign of weakness: giving into and acknowledging COVID risk. Trump had rarely been seen wearing a mask in his public engagements, even after the USA became the global leader in numbers of COVID infections and deaths. In the first presidential debate, only days prior to his announcement that he was infected with COVID, Trump had openly ridiculed his opponent in the imminent presidential election, Joe Biden, for wearing masks as a matter of course. Neither Trump nor most of his family members attending the debate wore masks. When Trump left hospital after treatment for COVID, he removed his mask as soon as he arrived at the White House to pose for photographers, despite still being contagious. Joe Biden, for his part, continued to wear his mask assiduously in his public appearances, pointedly commenting that 'Masks matter. These masks, they matter. It matters. It saves lives. It prevents the spread of disease' (Smith, 2020). In response to Trump's position on masks, memes circulating on social media included satirical cartoons showing Trump's open mouth blasting out coronavirus particles as he shouted, 'Real men don't wear masks!'. *The New Yorker* magazine's 14 October 2020 edition featured a cover with a drawing of a yelling Trump blinded by a face mask worn across his eyes. Meanwhile, entrepreneurial face mask producers marketed a mask showing Trump's unmistakable sneering mouth.

The position of Trump and Biden on face masks could not be more different: and they exemplify the deep political divisions in the USA on masks as a preventive measure against infection with the novel coronavirus. In this book, we discuss these, and all the manifold other complexities – and in some cases, contradictions – of the COVID face mask as a sociomaterial object. At first glance, the face mask recommended for everyday wear as a protective device against the spread of the coronavirus is an innocuous, low-tech object. Manufactured from either non-woven plastic or cloth materials, the COVID mask is designed

https://doi.org/10.1515/9783110723717-002

to be worn across the nose and mouth to reduce the exhalation and inhalation of viral particles in fine liquid droplets and miniscule particles (aerosols) that are propelled through the air via people's breath: particularly if they are speaking, shouting, sneezing, coughing, singing or breathing heavily due to physical exercise. Until the COVID pandemic erupted in early 2020, face masks were mostly worn by healthcare professionals when they were carrying out surgical or dental procedures, or by people in East Asian countries when they were out in public during the influenza or hay fever season. The emergence and rapid expansion across the globe of the novel coronavirus and the disease it causes, COVID-19, was officially declared by the World Health Organization (WHO) as a pandemic on 11 March 2020 (World Health Organization, 2020a). Since then, the coronavirus has spread rapidly, with many countries experiencing both first and second waves of the pandemic within six months of the first cases being identified in the city of Wuhan, China, at the end of 2019.

The COVID crisis has sparked new forms of sociality, everyday practices and ways of moving through time and space (Watson et al., 2020; Gammon and Ramshaw, 2020; Lupton and Willis, 2021). Among many other changes to private and public life, the COVID crisis has brought the humble face mask into new prominence. In the post-COVID world, it has become a significant object, positioned as one of the most important ways that people can protect themselves and others from infection with the novel coronavirus by acting as a barrier (however imperfect) between their breath and that of others. However, as our personal experiences recounted in the Preface and the furore around Trump's behaviour demonstrate, the face mask in the age of COVID is far more than a simple medical device. It has become the key symbol of the COVID crisis: with image after image of people's faces complete with masks in everyday settings used in popular culture to signify life in the COVID age (see, for example, Figure 1.1). The COVID mask is rich with symbolic meaning, affective forces and embodied sensations as well as practical value in these times of uncertainty, isolation, illness and death. The COVID mask is simultaneously a medical, social and multi-sensory device. Its presence or absence on the human face bears with it cultural, political and moral meanings. As the COVID crisis has intensified, fluctuated and diversified, so too, have these meanings.

The COVID crisis and the COVID face mask have a contemporary shared history and future, but the face mask's histories and politics pre-empt those of the novel coronavirus by centuries and go well beyond the frames of the health and medical context. Helped along by not only by Donald Trump's pronouncements and their accompanying media commentary but also by the statements of health officials, politicians and other public figures internationally, face masks have received greater prominence in public debates and popular culture than at any

Figure 1.1: A couple in New York City wearing COVID masks. Photo credit: Julian Wan, Unsplash.

other time in human history. There has been considerable debate in the medical and public health literature, policy circles, government agencies and the mass media about the value of masking wearing as a preventive practice. These debates have in some regions and countries flared into political activism and unrest, with groups agitating against mask wearing and particularly any attempts on the part of health authorities and government agencies to mandate their use. The most notable example is the USA, where protests erupted against face mask wearing that rest on 'sovereign individualism', a notion which is highly specific to the contemporary political climate in that country. In the USA, face masks have also been worn to make political statements: bearing anti-racist statements, for example, but also support for Trump.

Meanwhile, images and repurposings of the face mask have proliferated in popular culture, dominating social media discussions and generating an outpouring of internet memes, tweets, Instagram photos and TikTok videos that make fun of, support or agitate against mask wearing. Assisted by its prominent position on people's faces (or indeed, its equally obvious absence), the COVID mask has become a way of signifying the wearer's individuality, sense of style and beliefs or their ethical stance in relation to the need to protect their own and others' health. Celebrities and social media influencers have advocated for or against mask wearing as part of their branding. Museums, fashion designers, novelty fashion manufacturers and craftspeople on Etsy have identified the opportunity to profit from this sudden new market. A plethora of jokey masks can be found, with images of bizarre human facial features or animal faces, or warning people that 'If you can read this, you're too close' or 'Quit staring. Where's yours?'. Artists have played with the idea of the COVID mask by imagining incongruous objects as masks, creating highly embellished or stylised versions or using disposable masks as materials in their art. Memes, street art and other imagery have show popular cultural icons such as Baby Yoda and *Star Trek*'s Mr Spock with medical masks superimposed on their faces (see Figure 1.2), while subjects of famous portraits, including the Mona Lisa, the Girl with a Pearl Earring and a self-portrait by Vincent van Gogh have been similarly altered. Television dramas made post-COVID, including US series *Law & Order: SVU* and *Grey's Anatomy*, have begun to show characters wearing protective masks.

This chapter introduces the rationale for the book, addressing the question of why sociomaterial theories are so important to make sense of the meanings and practices related to the COVID face mask. It provides the context for understanding the COVID mask as a sociocultural artefact, discussing the history of the face mask internationally. We also provide an overview of the sociomaterial theoretical perspectives we are using in our analysis. Sociomaterialism is a broad term that is generally employed to described contemporary theoretical approaches interested in exploring the materialities of human existence. These approaches often have a strong political and ethical bent, directing attention to the ways in which humans and nonhumans live with and in relation to each other (Fox and Alldred, 2017). In this book, we draw particularly on Foucauldian theory, domestication theory and the more-than-human theory offered in Indigenous and First Nations philosophies and in the work of feminist new materialist scholars. We 'think with theory' (Jackson and Mazzei, 2012) to consider how the COVID

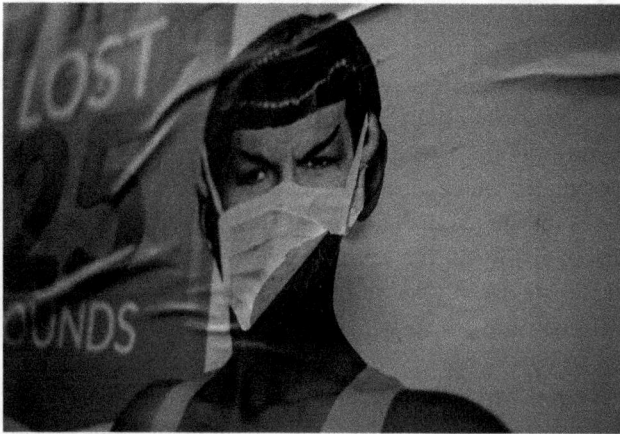

Figure 1.2: Street art featuring Mr Spock sporting a COVID mask. Photo credit: Nick Bolton, Unsplash.

mask has taken on extraordinary meanings, values and affective intensities. This chapter, therefore, provides the basis for elucidating the divergent sociocultural responses to COVID masks in contemporary political and geographical contexts that we discuss throughout the book.

Sociomaterialism: theoretical perspectives and concepts

In many strands of sociomaterialism theory, the concept of the assemblage is adopted to encapsulate the idea of ever-changing human-nonhuman gatherings. From a sociomaterialist perspective, the human body is conceptualised as porous and multiple, shape shifting, mobile and heterogeneous. It responds to its environment in ways that may increase or limit its agential capacities (Fox and Alldred, 2017; Andrews and Duff, 2019; Lupton, 2019, 2020a). Sociomaterialism perspectives position material objects such as face masks as contributing to assemblages of people with nonhuman things. It is with and through these combinations of humans and nonhumans that agencies and forces are generated. Thus, for example, we can think of the human body as a more-than-human assemblage of microbial flora, chemicals, fluids and other physical matter, ingesting certain substances, inhaling and exhaling gases, clothed in various objects, moving through a variety of places and spaces, continually interacting with other people and with other living creatures and using technologies such as spectacles, vehicles and smartphones ... just to name some of the potential agents in a more-than-human assemblage. Material objects such as face masks are designed with specific imagined uses in mind – or affordances. Human bodies too have affordances: the fleshy capacity to speak, sense, move, feel, respond, remember, exert awareness of and comment on their surroundings. The affordances of human bodies come together with the affordances of things in complex and multiple ways (Lupton, 2019, 2020a).

Michel Foucault's writings on biopolitics, biopower and governmentality identified how people's bodies are governed and managed by their incorporation of health-related advice into their everyday practices. Foucauldian analyses of medicine and public health highlight the intersections between the generation of medical knowledge, the role of health promotion and monitoring agencies and the emphasis on self-responsibility for maintaining good health that characterise contemporary western societies. In his books *The Birth of the Clinic* (1975), *Madness and Civilization* (1965) and *Discipline and Punish* (1977), Foucault reflected on the ways that certain forms of architecture – the medical clinic, the asylum, the prison – operate to discipline and monitor human bodies. In his later work, Foucault (1984, 1986) focused on the embodied practices involved in the care of the self and the government of public health. He traced modern states' preoccupation with managing the bodies and health of citizens back to the eighteenth century in Europe. A new discourse emerged at that time which positioned health as a collective concern that required population-based measures. By the late nineteenth century, the field of epidemiology had emerged, directed at monitoring population health and identifying causes of disease to better target preventive measures. Sociologists drawing on Foucault's scholarship have identified how medical and public health authorities were charged with gathering data to inform population health strategies: including information campaigns about the strategies people should use to protect themselves from infectious diseases. Dominant concepts of health and illness position many diseases as the result of a lack of taking responsibility for self-care. People are expected to take steps to manage their health status as entrepreneurial and self-disciplined subjects: engaging in practices such as regular exercise, body weight control, eating according to nutritional guidelines and avoiding tobacco and excessive alcohol consumption (Lupton, 1995, 2012; Petersen and Lupton, 1996).

As Foucauldian analyses of medicine and public health have demonstrated, late modern western concepts of human embodiment position the autonomous, closed-off individual body as the ideal (Lupton, 1995, 2012; Petersen and Lupton, 1996). Underpinning this ideal are deeply seated cultural anxieties about loss of bodily control and blurred boundaries between one's own body and those of others. Infectious disease outbreaks represent a strong challenge to these western notions of embodiment. People are confronted with the risk of a virus that moves easily between bodies through the air: contained in tiny droplets that are invisible to the human eye. Feminist philosophers such as Margrit Shildrick (1997) and Elizabeth Grosz (1994) have been writing for several decades about the implications of the ontological permeability of human bodies for the marginalisation and stigmatisation of bodies that are deemed to be 'too open' to the world. Shildrick's 'leaky bodies' and Grosz's 'volatile bodies' both emphasise this openness and fluidity of body boundaries. They point out that women's bodies in western cultures have traditionally been positioned as more fluid and permeable than men's bodies, as have the bodies of other marginalised social groups, such as people with disabilities, people of colour and members of the lower class and the poor. Privileged adult able-bodied white male bodies, in contrast, have been represented as better able to achieve the ideal of the tightly contained, highly disciplined and regulated body that has emerged in western cultures (Grosz, 1994; Lupton, 2012; Shildrick, 1997).

Domestication theory, taken up in media and cultural studies and in science and technology studies, has directed some attention to these aspects of quotidian lives. Domestication scholars argue that the process by which an object becomes domesticated involves four elements: appropriation, objectification, incorporation and conversion (Silverstone et al., 2005). Through appropriation, the individual or household acquires an object (usually by purchasing it), takes it home and gradually makes it part of their embodied practices or spaces. The object is thereby rendered from an impersonal to a personalised and customised thing. Objectification involves the location and arrangement of the object within the home, with the concept of 'home' being somewhat extended to intimate objects carried by or worn on the body of users, such as jewellery, clothing or smartphones. Incorporation refers to all the ways that the object is integrated into the user's daily life through frequent use, both intended and unintended, as well as the different capacities these uses generate. Finally, conversion refers to how the users of the object integrate the object into their sense of self identity and use the object to express themselves (Silverstone et al., 2005). From this perspective, the household is a moral economy involved in the dynamic production and exchange of meanings with the public world, and the domestication of the object expresses meanings about the household to broader society. The adoption of objects and technologies, according to the domestication approach, is negotiated through this moral economy, rather than determined purely by the objects or the people who use them. As Berker and colleagues (2006: 2) argue: 'technologies have to be house-trained, they have to be integrated into the structures, daily routines and values of users and their environments'. In this book, we show how COVID masks can become domesticated (or 'house-trained') – but also how some people resist domestication of the mask.

While domestication theory and Foucault's work identified integral material elements that contribute to knowledges and practices related to human bodies, the sociomaterialist scholarship we characterise as 'more-than-human theory' offers a more expansive understanding. More-than-human theory is espoused in the work of Indigenous and First Nations philosophies (Watts, 2013; TallBear, 2014; Bawaka Country et al., 2016; Todd, 2016; Hernández et al., 2020) and that of feminist new materialism scholars such as Donna Haraway (2016), Karen Barad (2007), Jane Bennett (2009), Rosi Braidotti (2019), Stacey Alaimo (2016) and Maria Puig de la Bellacasa (2017). The term 'more-than' term vastly expands the category and scope of the human, acknowledging that people are always already more-than-human, coming together and living with and through other people, other animals and living creatures, and with non-living things, places and spaces (Lupton, 2019, 2020a). It emphasises the relational and distributed nature of agencies when assemblages of humans and nonhumans come together, and the forces that are generated by these gatherings.

From this perspective, humans and nonhumans alone do not possess or exert agency: it is only in their assemblages with other actors that agency is generated and expressed. Bennett's concept of 'thing-power' is an evocative way of encapsulating the vibrancies and vitalities of these assemblages; or as she puts it, 'the curious ability of inanimate things to animate, to act, to product effects dramatic and subtle' (2004: 351). Bennett (2001) also draws attention to what she describes as the 'enchantments' generated with and through people's encounters with nonhuman things; the powerful affective intensities that animate people's investments in and attachments to objects. Barad (2007) uses the term 'intra-act' rather than 'interact' to describe these forces as a way of highlighting their emergent and distributed nature, and how 'matter makes itself felt' (Barad in Dolphijn and Van der Tuin, 2012: 59). This is a post-Foucauldian approach to power that deepens his insights about the productive and vital nature of power by acknowledging the productive and vital forces of more-than-human assemblages.

As it has been espoused by western philosophers, more-than-human theory has been subject to some criticism for ignoring insights from Indigenous and First Nations philosophies and not devoting enough attention to sociodemographic inequalities and power asymmetries, the potential destructive or harmful capacities and agencies that may be generated as part of thing-power or the ways in which these capacities and agencies can be used for political change (Todd, 2016; Hernández et al., 2020). There is no reason, however, why more-than-human theory cannot be taken up to examine these kinds of asymmetries as part of an onto-ethico-epistemological position (Barad, 2007; Braidotti, 2019) that helps us to 'see, think and feel differently' (Renold, 2018: 39). For more-than-human theorists drawing on Indigenous and First Nations perspectives as well as on feminist new materialism, categories of the human always involve questions of power, privilege, colonialism, racialisation, social discrimination and marginalisation. Social theorists and researchers who have engaged with more-than-human theory emphasise the importance of moving beyond the social structural ('macro') elements of power relations and social inequalities that preoccupied earlier materialism (particularly influenced by political economy theory) and the linguistic emphasis in post-structuralist approaches. Focusing more attention on the micropolitics and affective dimensions of the relations and encounters of humans not only with each other but with nonhumans can identify the dynamics and complexities of health and wellbeing.

From this perspective, phenomena such as 'health' are defined as the capacity of bodies to affect and be affected, focusing on what bodies can do (Andrews and Duff, 2019; Lupton, 2019, 2020a). Indeed, a close emphasis on the details of human-nonhuman assemblages and the ways in which capacities can be opened or closed off can lead to explorations of how features of people's lives such as their relative socioeconomic advantage, their living circumstances (for example, the quality of their housing, food intake, access to high-quality education, safe and consistent employment and other resources) and the other life experiences that are part of their assemblages with nonhumans (Renold, 2018; Andrews and Duff, 2019; Braidotti, 2019; Lupton, 2019, 2020a). As Braidotti (2020) argues, in this age of global crises involving complex assemblages of people with other living things, more-than-human theory and inquiry can contribute to an affirmative and relational ethics in which all forms of life are recognised as important, intimate connections between humans and nonhumans are understood, and the agencies and capacities that are distributed between people and other things can be identified. The situated, emergent and temporal nature of these phenomena require constant attention. Haraway (1988) points out that a standpoint is always 'from somewhere' and we need to identify the context. For Barad (2003) such an approach can demonstrate 'how matter comes to matter'.

This more-than-human perspective has been central to Indigenous and First Nation cosmologies for millennia (Watts, 2013; TallBear, 2014; Bawaka Country et al., 2016; Todd, 2016; Hernández et al., 2020). However, it has only recently been recognised as important in the post-Enlightenment societies of the Global North and recuperated into western social theory and inquiry: predominantly because of the growing recognition that Anthropogenic climate change and environmental degradation has become a global crisis (Alaimo, 2016; Haraway, 2016; Braidotti, 2019). Humans can no longer ignore their role in contributing to this devastation, nor that they are imbricated within more-than-human ecologies and are therefore harmed by it. In the wake of the COVID crisis, this emphasis on human and nonhuman encounters has become even more pressing (Braidotti, 2020). The COVID-19 crisis has brought the more-than-human dimensions of life into sharp relief: including many existing as well as new social inequities (Bambra et al., 2020; Braidotti, 2020; Lupton and Willis, 2021; Lupton, 2020b).

The novel coronavirus entity, transferred to humans via a nonhuman animal vector in the context of a changing climate and food supply ecosystem, has spread rapidly around the world due to the movements of human bodies facilitated by transport systems and people's interactions with other people. Measures to limit the spread of COVID involve complex assemblages of human bodies with other people and with things such as soap and water, face masks, COVID tests, thermometers and potential vaccines. Places and spaces have been deemed either risky (public spaces) or safe (the home) in terms of the risk of contagion; either to be avoided or a place of refuge. And as the histories of previous pandemics have also demonstrated, the COVID crisis has shown how some people and social groups are considered to be as less 'human' (older people in care homes, people with disabilities, people of colour, homeless people, the poor, people living in the Global South), and therefore, as less important or worthy of care or protection, than others (Lupton, 2021; Bambra et al., 2020; Braidotti, 2020). It is in understanding how these assemblages come together and what agencies and capacities for action are generated with and through these gatherings and practices that a sociomaterial approach offers a way forward.

The history of the face mask

Historical accounts of face mask use demonstrate the significant cultural differences in understandings and practices related to mass masking for health purposes. Face coverings have been traditionally worn in some countries for reasons other than health concerns. Full or partial face coverings worn by women, such as veiling, is common practice in regions such as the Middle East for religious and modesty reasons but also often as a fashionable practice of self-enhancement (Crăciun, 2017). In countries such as Japan, China, Korea and other East Asian nations, face mask wearing has a long tradition (Burgess and Horii, 2012; Horii, 2014; Ma and Zhan, 2020). Disposable medical masks are commonly worn in the winter months to protect against contracting and spreading infectious disease such as colds and influenza, and at other times of the year as a way of mitigating exposure to pollens or air pollution (Horii, 2014; Ma and Zhan, 2020). Chinese people have been encouraged by public health authorities to wear protective masks during epidemics from the early 1900s, when China contended with outbreaks of the plague, Spanish influenza, cholera, smallpox and other serious infectious diseases (Lynteris, 2018; Wei, 2020). In Japan, people have worn medical masks largely out of habit: as a form of health ritual but also, in the case of women, to protect their faces against the sun (Burgess and Horii, 2012). The outbreak of SARS (Severe Acute Respiratory Syndrome) in 2002 and avian influenza between 2004 and 2006, both predominantly affecting East Asian countries, further promoted mask wearing for health reasons (Burgess and Horii, 2012; Ma and Zhan, 2020). The swine flu pandemic in 2009 also spurred greater use of face masks in Japan, following public health campaigns recommending their use (Burgess and Horii, 2012), as did the Fukushima nuclear plant accident in 2011 (Horii, 2014). For Japanese people pre-COVID therefore, mask wearing became a taken-for-granted practice that helped them deal with and gain a sense of control over diffuse invisible threats and uncertainties (Horii, 2014).

In most countries in the Global North, pre-COVID most people who were not healthcare workers had little or no experience of face masks worn as an everyday preventive health measure. They have had to learn how to make sense of face masking as a protective practice and how to incorporate face masks into their everyday practices and routines. In Europe, the use of masks for the purpose of preventing against contracting infectious disease emerged as part of a range of preventive measures, changing in their appearance and function as ideas about disease causation have changed. In early modern times, it was believed that foul air was full of contagion ('miasma') and breathing in these odours would cause illness. People attempted to avoid the most odorous places as a preventive health measure, or carried sweet-smelling herbs to breathe in (Cole, 2010). With the advent of scientific medicine in the late nineteenth century, the discovery of the existence of microorganisms and the development of the germ theory of disease, the causes of infectious disease and its spread became better understood. By the turn of the twentieth century, methods of containing infectious disease had begun to include hygienic measures such as using soap and water for hand washing to remove microbes (Pelling, 1993). It became recognised that human bodies in close contact with each other could spread disease between themselves via microbial contamination. While quarantine and physical isolation strategies had been in place as measures of containment since the fourteenth century (Bashford, 2016), it was now understood that the reason these strategies worked – and why face masks could also provide a barrier to infection – was because they reduced the opportunity for microbes to be conveyed from person to person (Matuschek et al., 2020).

Surgeons began to wear cotton gauze face masks in the USA and Germany in the 1920s to reduce infection in patients during surgical procedures, but this practice only began to become common procedure in these countries and elsewhere in the 1940s (Matuschek et al., 2020). The Spanish influenza pandemic of 1918–1919 was a watershed occasion for everyday masking as a preventive health measure: including greater use of masks in East Asia (Burgess and Horii, 2012; Horii, 2014) but even in western countries where masks had not commonly been worn for health reasons (Flaskerud, 2020). People across the globe were encouraged to make their own fabric masks and wear them as a protective measure, as shown in Figure 1.3, a poster used in a health campaign in the province of Alberta, Canada. The poster provides instructions for how to make a simple mask from 'ordinary cheesecloth' by folding it and attaching cords to tie it on the head.

Medical and public health advice on mass masking in COVID times

Early in the COVID pandemic, debate erupted among medical and public health authorities as to how effective face masks would be as a preventive measure for

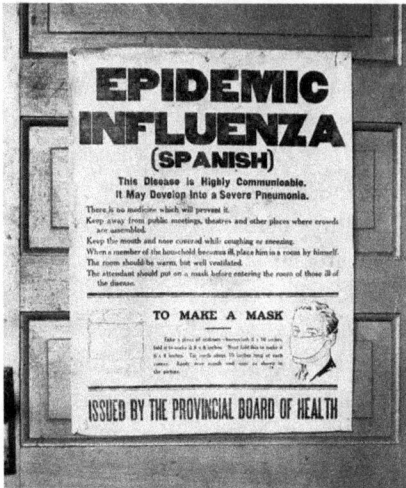

Figure 1.3: Poster issued by the Provincial Board of Health, Alberta, Canada. Photo credit: Glenbow Museum.

the general public: sometimes referred to in the medical literature as 'mass' or 'universal' masking. Face masking was not as simple a public health preventive measure as it may have first appeared. Very few studies had been published at that point in the medical literature about the efficacy of face mask wearing to limit the spread of an infectious agent such as SARS-CoV-2, with no research on the usefulness of cloth masks for the public (Greenhalgh et al., 2020). In the early months of COVID, there was some confusion in the medical literature about whether masks should be worn as a way of protecting others from infection or to protect the mask wearer from others. Several rapid reviews were published in the medical literature suggesting that mass masking was not an effective preventive measure against coronavirus infection for those wearing them (Martin et al., 2020), partly because it was acknowledged that masks could not block out all virus-containing droplets and aerosols (Cheng et al., 2020; Greenhalgh et al., 2020). However, medical and public health authorities then began to focus on the mask's role in protecting others from the wearer, if that person happened to be infected with the novel coronavirus. As Cheng and colleagues (2020) put it in the title of their piece published in *The Lancet*, wearing face masks demonstrated 'altruism and solidarity'.

Mask mandates, regulations and recommendations changed often in the early months of the novel coronavirus outbreak. There was concern in the early months that there would not be enough medical masks for healthcare workers, particularly if members of the public started to wear them regularly (Cheng et al., 2020). On 3 March, the WHO released a statement emphasising the shortage of personal protective equipment (PPE) worldwide, including medical masks (World Health Organization, 2020b). Even after officially declaring COVID-19 as a pandemic, the WHO did not initially recommend that people without COVID symptoms outside healthcare settings should wear a mask in public places. Concern was also expressed in the medical and public health literature that people using masks in community rather than healthcare settings may not wear or handle them appropriately and could therefore increase their chance of infection by touching the outside of the mask after wear or not disposing of it or cleaning it properly if it was a reusable mask (Greenhalgh et al., 2020; Martin et al., 2020). Another potential problem expressed with advocating widespread mask wearing was that it might give people a false sense of security or complacency and therefore dissuade them from other important preventive measures such as regular handwashing and physical distancing (Cheng et al., 2020; Martin et al., 2020). It was also suggested that mass masking might lead to a 'climate of fear' by reinforcing the seriousness of the COVID crisis when people saw others routinely wearing masks (Martin et al., 2020).

In what was a rapidly changing policy environment and as more medical knowledge about SARS-CoV-2 and COVID-19 was developed, by April 2020 there was evidence of a shift in advice (Javid et al., 2020). The emphasis in public health messaging was the need for those people with COVID symptoms to wear masks in public places. By June, WHO had begun to recommend mass masking, reversing its previous advice (Martin et al., 2020). As medical research began to demonstrate that people could be infected with COVID but display few, or no, symptoms, some recommendations about mask wearing in some countries began to acknowledge the risk of 'asymptomatic shedders' of the virus (Cheng et al., 2020). Venezuela, Vietnam, the Czech Republic, Slovakia, Bosnia and Herzegovina all mandated mask use in public areas in March 2020, with Colombia, the United Arab Emirates, Cuba, Ecuador, Austria, Morocco, Turkey and several other countries, including numerous African countries, following in April (Anonymous, 2020a).

By July 2020, over one hundred countries had introduced policies requiring their citizens to use face coverings in situations where physical distancing is not possible, such as crowded places and public transport (Flaskerud, 2020). At this point in the pandemic, in countries that were recommending high levels of physical isolation and staying at home for those with COVID symptoms, face masks were not seen as necessary (Kayat, 2020). Many East Asian countries continued with their usual practices of wearing masks to prevent spreading infectious disease, supported by public health messaging about COVID risk. For example, China introduced mandatory mask wearing in some areas early on in the pandemic but relaxed these regulations towards the middle of 2020 as case numbers dropped (Hernández, 2020). South Korea introduced a mask mandate in metropolitan areas only during its second wave of COVID in August 2020, after very successfully combating the first wave without mandating masks. However, in that country, mask wearing was widely adopted even if not required (Smith and Cha, 2020). In Japan, mask wearing was already widespread due to cultural familiarity with the practice and the government issuing every household with two reusable cloth masks (Anonymous, 2020b).

Anglophone nations for the most part avoided mandating mass making at a national level in the early months of the pandemic and indeed at first attempted to dissuade rather than encourage public use (Javid et al., 2020). For example, the US Surgeon General, Jerome Adams, initially repeatedly warned US citizens not to wear masks and advised that they should rely on washing their hands and avoiding crowds. His advice could not be less equivocal. In March 2020, Adams tweeted the following statement: 'Seriously people – STOP BUYING MASKS! They are NOT effective in preventing general public from catching #Coronavirus, but if healthcare providers can't get them to care for sick patients, it puts them and our communities at risk!' (Zhou et al., 2020). The influential Centers for Disease Control and Prevention (CDC) in the USA was at first equivocal, stating that masks should be worn only by American with COVID symptoms and not the general public (Greenhalgh et al., 2020). In April 2020, however, its advice had changed. By August 2020, on its webpage devoted to COVID information the CDC recommended that face masks could 'help slow the spread of COVID-19' and that 'people should wear masks in public settings and when around people who don't live in your household, especially when other social distancing measures are difficult to maintain' (Centers for Disease Control and Prevention, 2020a).

Across the Global North, signs began to appear in public places to notify people they should wear a mask, as shown in Figure 1.4. By November 2020, when several US regions were experiencing massive surges in COVID infections, the authorities of more than 20 states had begun to require mask wearing in public (#Masks4All, 2020b). In Canada, the Chief Public Health Officer, Theresa Tam, in August 2020 described masks as an 'added layer of protection' to use when physical distancing was difficult (Harris, 2020). Australian authorities at first recommended that masks should not be worn, declaring them to have minimal protection against acquiring infection with the coronavirus. However, by the time a second wave of COVID occurred in mid-2020 in the state of Victoria, mass masking was eventually required as one of the key preventive measures, along with a re-imposed lockdown (Australian Government Department of Health, 2020b).

Once it was largely agreed that mask wearing could offer a way of limiting the spread of the coronavirus in the community, writers in medical and public health journals began to cover the complexities of using the 'right' type of

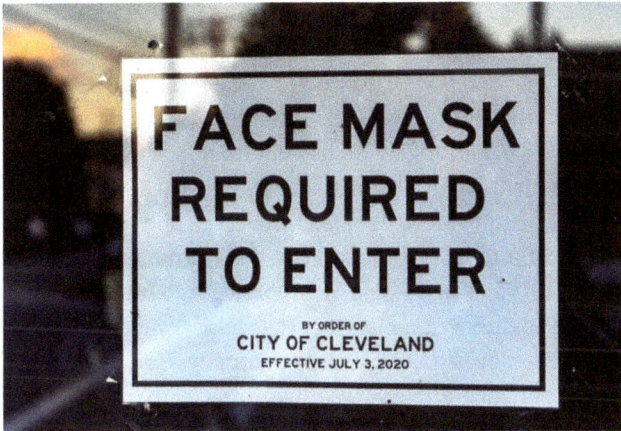

Figure 1.4: Notice that face masks should be worn inside this building, City of Cleveland, USA. Photo credit: DJ Johnson, Unsplash.

mask and wearing it in the 'right' way. It was recognised that the level of protection varied with the type of face covering chosen: some materials were far better than others in limiting the escape of droplets and aerosols from the wearer's nose and mouth and fitted masks were more protective than loose, scarf-like coverings (Greenhalgh et al., 2020; Javid et al., 2020). People using face coverings required expert advice as to how to don and doff them carefully to avoid contamination from handling them and how best to store soiled re-usable masks and wash them once worn. By December 2020, the WHO was advising people living in a community in which COVID was spreading to 'Make wearing a mask a normal part of being around people' and that 'The appropriate use, storage and cleaning or disposal of masks are essential to make them as effective as possible' (World Health Organization, 2020c). Despite this growing consensus globally by medical and public health authorities and governments, in December 2020 there were still several countries who had not yet mandated mass masking in public places nationally. These included countries with very high rates of COVID infections and deaths such as USA (where only some states required mask wearing), Russia and Brazil as well as countries such as Japan, India, South Korea and China where voluntary mass wearing was already very high among the community (#Masks4All, 2020b).

Rest of the book

In the remaining chapters of this book, each addresses a discrete topic related to the sociomaterial dimensions of COVID face masks. Chapter 2 focuses on micro-political and macropolitical aspects, ranging across international disputes over medical mask production and supply, the role played by peak health organisations such as the WHO, mass media and social media coverage and social movements seeking both to support and agitate against mass masking. In Chapter 3, we address the ways that COVID masks become incorporated into human bodies and everyday practices, bringing domestication theory together with more-than-human perspectives. Chapter 4 moves us deeper into our analysis of the embodied sensory and affective experience of mask wearing, focusing particularly on breath and breathing with and through a COVID mask. The artefact of the hand-crafted COVID mask is examined in Chapter 5. We bring perspectives from social analyses of Do-It-Yourself (DIY) and crafting cultures to discuss the sociomaterialities of four different kinds of hand-crafted masks: the artisan mask, the home-made mask, the makeshift mask and the community drive mask. In Chapter 6, we turn our attention towards the concept of care and how this may be applied not only in the context of medical care and caring for oneself or other people by wearing a COVID mask, but the implications for the environment of careless use and disposal of masks. The Epilogue brings together the threads of our arguments and provides some final thoughts on the COVID mask as a sociomaterial phenomenon.

2 Face Mask Politics

Introduction

As face masks have come to be a key symbol of the COVID crisis, they have also become deeply politicised, especially in the USA. The COVID crisis as an 'event' has been continually shifting, providing a disorienting landscape in which to situate critical debates around the face mask. In this chapter, we orient the emerging political tensions surrounding face masks by tracing how they rose to public awareness through pivotal events such as the WHO's change in public health policy on mass masking, and growing supply-chain concerns around masks, as well media discourses surrounding these events. We examine how masks came to be a key site of political division and activism during the pandemic, even being used to signal alliance to a particular political party or social movement. As masks became a central focus of COVID control and prevention and with the introduction of mask mandates or recommendations across the world, the anti-mask movement emerged. Countering this movement were discussions organised around the hashtag #Masks4All, which argued in favour of mask-wearing as a primary prevention strategy. These events range in scope from international political conflicts at the macro-level to micro-level shifts in daily life.

Medical mask supply politics

Even before the COVID crisis had made a dramatic impact on everyday lives, it had already significantly disrupted the global supply chains through which disposable medical face masks, face piece respirators and other PPE used by healthcare workers – such as gowns, gloves and eye protection – are procured. Medical masks are designed specifically for use in clinical settings, with respirators reserved for healthcare professionals dealing with patients in situations where aerosol spread of SARS-CoV-2 is likely. It is usually recommended that publics outside these settings use non-medical fabric masks instead, unless they are in vulnerable groups or caring for a person infected with the coronavirus (Centers for Disease Control and Prevention, 2020a). As demand increased for PPE once the pandemic had taken hold, many countries also introduced restrictions on the export of products like disposable medical masks to preserve domestic stock (OECD, 2020a). China is the largest manufacturer of medical masks, with half of the global market share (OECD, 2020b). However, given

https://doi.org/10.1515/9783110723717-003

that China was also the country most affected by the spread of the novel coronavirus in the early months, its domestic need for medical masks for healthcare workers rapidly exceeded their production, and imported masks were needed to supplement its stocks (Gereffi, 2020).

Despite increasing mask production ten-fold by March 2020, China was still struggling to meet domestic and international demand. Although there was no export ban in place, the Chinese government had purchased all medical masks produced in January and February 2020, with mask export resuming in March (OECD, 2020a). The flow-on political effects of medical mask shortages have been significant. Tensions increasing between China and the US over trade agreements needed to ensure the supply chains remain open were exacerbated by President Donald Trump fanning the flames of racism by referring to the novel coronavirus as the 'Chinese virus' (Bradsher and Swanson, 2020). Furthermore, news outlets and human rights organisations began reporting disturbing details of Chinese face mask factories allegedly exploiting or even coercing minority groups such as the Muslim Uighurs in modern slavery conditions to ramp up production for foreign export (Mathers, 2020).

At the same time, 'mask diplomacy' has also been observed. One example is Japan's donation of over a million medical masks to China, which Brookings Institution scholars contend has had a significant positive impact on the deep-seated hostility between the two nations (Li and McElveen, 2020). As their production capacity grew and government regulations loosened, China too began donating medical masks internationally to Italy, Spain and South Korea (Zhou, 2020). This action in turn further inflamed tensions with the US government, who were suspicious that these donations could be Chinese government propaganda (Wong and Mozur, 2020). Unsurprisingly, low-income nations have been significantly disadvantaged from the outset, with a recent study (McMahon et al., 2020) identifying that even before the pandemic many were already understocked on the key PPE necessary to equip medical personnel during the pandemic. These inequalities have been exacerbated by high income nations drawing on their vast resources, with accusations of 'modern piracy' involving PPE being sold to the highest bidder rather than those most in need (Bradley, 2020).

In early February 2020, the WHO Director-General Dr Tedros Adhanom Ghebreyesus addressed the issue, stating that stockpiles of PPE were depleted and '[d]emand is up to 100 times higher than normal, and prices are up to 20 times higher. This situation has been exacerbated by widespread, inappropriate use of PPE outside patient care' (World Health Organization, 2020d). A month later, the WHO (2020f) urged governments and industry to increase the production of PPE by forty per cent, and criticised 'panic buying, hoarding and misuse'. These statements draw a connection between the macro-political level issues the movement of COVID masks brings about, and micro-level practices. Behaviour such as panic buying and hoarding have been problems in many countries throughout the COVID crisis, often generating interpersonal conflict in supermarkets over low stocks of toilet paper, hand sanitiser, food essentials and other everyday items (Guardian staff and agencies, 2020). High-profile cases of individuals hoarding thousands of masks and other PPE to sell at inflated prices have drawn widespread news coverage and social media outrage. The arrest of one hoarder in US who was stockpiling respirator masks intended for medical personnel was captured on a video that attracted nearly 3 million views (ABC News, 2020). A stockpiler in the UK was named and shamed by tabloid newspaper *The Daily Mail* after refusing to hand over £2,500 worth of PPE (Ridler, 2020). In April, multinational online retailer Amazon stopped selling respirators and medical masks to the public to attempt to address stockpiling and misuse of PPE (O'Kane, 2020).

Alongside these tensions surrounding stockpiling, the early months of the pandemic's spread (January to March) saw the beginning of technical language around protective equipment, different mask grades, qualities and materials integrating into the popular lexicon. Google Trends data shows that Google searches for 'PPE' began to increase rapidly in early March before reaching a peak in early April. Governing bodies and health authorities began to urge the public not to purchase medical respirator masks so as not to deplete already dwindling supplies that ought to be reserved for medical staff (Cramer and Sheikh, 2020; Patty, 2020). In doing so, the general population became more familiar with the differences between these higher-tech masks, ordinary medical masks and the growing array of cloth masks produced to meet the needs of everyday people. News outlets (Gunders, 2020; Mathew, 2020), as well as organisations like the Red Cross Australia (2020), published guides to help consumers identify different masks and regularly update them on masks that were no longer recommended.

Notably, masks with valves fitted were identified as offering limited protection (Verma et al., 2020). In addition, a study of different face masks concluded that neck gaiters and bandanas 'offer very little protection' (Fischer et al., 2020): a finding that led to news outlets globally reporting that these types of masks were 'worse' than no mask at all (Kelleher, 2020). In the days following, these reports were complicated by the clarification that the study was focused only on testing methods to test masks, rather than establishing definitive proof of which masks provided the best protection, and the authors of the study felt their findings had been misrepresented (Schive, 2020). That this study received high levels of publicity suggests that public appetite for information about mask wearing has become whetted in a context in which everyday consumers were beginning to view masks as a part of everyday life.

The WHO's public change in position on the face mask was a significant barometer of the unfolding global tensions surrounding this object that had become so much a part of the pandemic symbolically and politically, as well as in the embodied experience of ordinary people. On 5 June 2020, the WHO revised its previous position on the wearing of face masks. It had previously released a statement on 6 April affirming that 'wide use of masks by healthy people in the community setting is not supported by current evidence and carries uncertainties and critical risks' (World Health Organization, 2020e). The updated statement in June was a significant shift on this position, advising that all medical workers should be wearing medical masks continuously. The WHO further encouraged the widespread wearing of masks in places with community transmission of the novel coronavirus, where physical distancing is not always possible or where population vulnerability is higher (for example, older people and people with pre-existing medical conditions) (World Health Organization, 2020f).

Mask-wearing politics

The introduction of mask mandates or even mask recommendations cannot be understood outside of the context of these existing laws banning masks, demonstrating that the legal, social and cultural meanings of face masks continue to be negotiated and renegotiated. As public health advice began to conform globally to recommending mass masking, in some countries, most notably the USA, the COVID mask became a central object of political issues and debates surrounding the pandemic. In particular, questions about whether mask wearing ought to be widespread as a matter of public safety or whether it should be a personal choice on the grounds of individual freedom emerged as a critical battleground.

Mask wearing, however, has long been the source of political dissension and even legal sanctions. Many of the countries and US states introducing mask mandates have in pre-COVID times implemented laws that criminalised wearing masks. Women wearing face coverings for modesty or religious reasons in western countries have frequently been subjected to attack and racist abuse. Several European nations including France, Denmark, Austria, Bulgaria, Belgium and the Netherlands have passed laws in the past few years to ban face coverings worn in public places. It was claimed that public safety was at issue, but these laws are clearly motivated by anti-Islamic sentiment and a desire to ease the discomfort of difference through forced cultural homogeneity (Amir-Moazami, 2016). In the USA, laws had largely focused on preventing people gathering and wearing masks. They were used to arrest members of the Klu Klux Klan and, more recently, Occupy Wall Street protestors (Flaskerud, 2020). Hong Kong has controversially maintained a ban on face coverings at 'unlawful gatherings', a law that has been used during the recent protests in the city, even while masks are widely worn to reduce the risk of COVID spread (Davidson, 2020a).

In the context of the COVID crisis that had been building for months, tensions were running high in many countries as mask regulations continued to change, especially in places where such close regulation of citizens' behaviour was significantly out of the ordinary. It is perhaps not surprising then that as masks began to be recommended or mandated alongside other restrictions to address the transmission of the novel coronavirus in the community, there were protests in many countries as citizens expressed frustration. As shown in Figure 2.1, face masks were often at the centre of these debates: both because of their symbolic position as a part of the COVID pandemic as well as their role in shaping everyday embodied and sensory experiences of the pandemic through their wear. The emergence of global anti-mask movements oriented around personal freedom and opposition to government control has taken place predominantly in wealthy countries in the Global North, with anti-mask protests occurring in UK, Australia, Italy, Ireland, Germany and Canada, but in the USA above all (Philipose, 2020; Beresford, 2020). For those people who resist mask wearing, the sheer lack of a mask on their face is itself a protest. More creatively, some anti-maskers have taken to deliberately wearing masks made of ineffective materials such as mesh or lace, or sport masks with anti-mask messages (Segall, 2020). Examples of such messages include 'This mask is as useless as the government' and 'Only wearing this so I don't get fined'.

These protests share similarities with historical anti-mask movements that emerged in the United States during the 1918 influenza epidemic, where masks were made mandatory in some areas. Like the anti-mask and anti-lockdown down protests in COVID times, the 'Anti-Mask League' formed in 1919 was focused on individual freedom and liberty, as well as questioning the scientific legitimacy of claims that mask protect against influenza (Dolan, 2020). However, contemporary anti-mask protests in wealthy countries differ significantly from those that occurred in other nations that targeted poor governmental response to the COVID crisis. Such protests took place in countries such as India, where migrant workers were particularly badly affected by government policy (Puranam, 2020), Rwanda, where refugees were stranded by lockdown conditions (Associated Press, 2020), and Nigeria, where COVID patients protested against their poor treatment in government facilities (Anadolu Agency, 2020).

Journalist Emily Stewart (2020) conducted interviews with US-based 'anti-maskers' who offered explanations for their stance on face masks. Some people expressed strongly affective responses. They emphasised their frustration and discomfort with the act of mask wearing, and the anger they felt at being forced

Figure 2.1: A mask-shaped sign used at a protest in Trafalgar Square in London, UK. Photo credit: Ehimetalor Akhere Unuabona, Unsplash.

to wear one. Some anti-maskers Stewart interviewed did not object to mask wearing per se, but rather the mandates, emphasising personal choice and bodily autonomy. However, for many people, there was a significant connection between their objection to mask wearing and their other beliefs in conspiracy theories concerning the COVID pandemic: that it was a hoax or vastly overstated as a threat. Positioning themselves as able to see through the lies of health officials, these individuals were convinced that medical authorities' claims concerning the pandemic are really an attempt to create a more docile population and that masks are an important part of this strategy. They argued that the emphasis on the social good of wearing masks (that is to protect others, especially the vulnerable) is vastly overstated or outright fiction that serves only to encourage the submission of citizens to the state. As one interviewee explains: 'I'm laughing because grandmas and grandpas die all the time. It's sad. But here's the thing: It's about blind obedience and compliance' (Stewart, 2020).

In her interviews with anti-maskers, Stewart found that they tended to hold politically conservative views. This observation was mirrored in a July Gallup survey of Americans, which found that 98% of Democrats said they had worn a mask in public in the past week, compared with 66% of Republicans (Brenan, 2020). This finding can perhaps be attributed to President Donald Trump's long-time refusal to wear a face mask and mocking of rival Democratic candidate presidential Joe Biden for wearing one (Chapter 1). Trump proclaimed in June 2020 that he thought some people wore face masks to indicate their disapproval of him and expressed doubt that they were useful at all in preventing the spread of the coronavirus (Sheth, 2020). Interestingly, in neighbouring Canada, those people who objected to mandatory mask-wearing were more evenly spread across the political spectrum (Janzwood and Lee, 2020).

Globally, anti-maskers have also been associated with anti-government groups and with the 'sovereign citizen' movement, a loosely affiliated right-wing extremist group described by the US Federal Bureau of Investigations as terrorists (Sarteschi, 2020). According to criminologist Christine Sarteschi (2020), '[t]he core belief of sovereign citizens, is their proclaiming the American government to be illegitimate'. Proponents of this ideology do not believe themselves to be subject to any US government laws. Although the movement has its origins in the USA, it has spread significantly, with documented sovereign citizens in Australia, the UK, Canada, South Africa, Singapore and similar movements in Germany and Russia (Berger, 2016; Sarteschi, 2020).

Both left-ring and right-wing commentators have expressed further concern about the increased surveillance and excessive disciplining of people's public behaviours that can be part of mass masking initiatives. Pre-COVID, researchers in surveillance studies and new media studies have drawn attention to the contraventions of personal privacy involved in digitised forms of public surveillance, such as CCTV, drones and facial recognition systems (Lyon, 2018). Post-COVID, similar criticisms have been made of the potential for such systems to be employed to identify, shame or penalise people who are not wearing masks in public places in situations where there is little opportunity for such individuals to provide explanations for the lack of a mask or appeal a penalty. Some jurisdictions, including the city of Or Akiva, Israel, have introduced digital surveillance cameras to monitor whether citizens are complying with the government's COVID guidelines: including mask wearing in public. Those people identified as non-compliant are identified by PA systems announcing their transgressions, and law enforcement can be activated by the system (Anonymous, 2020c). In a range of other countries, including China, Italy, Spain and France, 'shout drones' equipped with digital cameras have been used by government and police authorities to identify people not complying with COVID restrictions and literally yell at them using loudspeakers on the drones (Greenwood, 2020). Facial recognition systems were initially disrupted by mass masking practices, but the vendors of the systems are quickly adapting their software to respond to these conditions so that they can 'see through' masks (Margit, 2020). All these developments can potentially lead to unfair discrimination, the enforcement of authoritarianism, the use of people's personal digitised information for political targeting, the flouting of human rights and civil liberties and what has been described as 'surveillance creep', or the extension of surveillance practices well beyond their initial rationale and application (Calvo et al., 2020; Couch et al., 2020).

Beyond support or dissent for mask wearing being affiliated with a political party, face masks have themselves been used to signal national and political affiliations and as a tool for political expression across all ideological spectrums. COVID masks have complemented other public displays of political allegiance, such as T-shirts and signs. For example, during the 2020 US presidential election, Trump supporters wore masks with the slogans 'Make America great again' or 'Trump 2020', while those advocating for LGBTQI rights have sported rainbow masks and environmental campaigners have worn appropriately themed masks. Protestors have also used masks as a way of drawing attention to their governments' mishandling of the COVID crisis, including in Serbia and Lebanon.

As Figure 2.2 shows, wearing masks with national flag themes became a common practice during 2020. Political leaders such as Australia's Prime Minister, Scott Morrison, have appeared in public wearing such masks. In Uganda, the Government distributed face masks were provided in colours that were associated with the ruling political party, a move that development studies scholar Innocent Anguyo (2020) describes as a way to 'tap into the potential of masks for political dressing'. In response, Ugandans have been mobilising the mask as a political statement, wearing the colour of the political party they support or wearing different coloured masks at different times to maximise personal gains. In Australia, face masks have also been made and sold by Indigenous

Figure 2.2: A mask used to display national identity. Photo credit: Bermix Studio, Unsplash.

groups featuring artwork from Aboriginal artists, offering consumers a way to visibly as well as economically support Indigenous businesses during the pandemic (Vrajlal, 2020).

In addition to being split along political lines, some scholars have argued that mask wearing is also a gendered practice. Conservative US television commentator Tomi Lahren recently criticised Joe Biden for wearing a face mask. In a tweet she suggested, 'Might as well carry a purse with that mask, Joe': a clear suggestion that mask wearing is a feminine or effeminate practice (Dicker, 2020). A recent US survey exploring masculinity and mask-wearing found that 'masculine toughness is consistently related to higher negative feelings and lower positive feelings about mask wearing' (Palmer and Peterson, 2020: 3). A similar survey, also in the US, further found that men are more likely than women to agree that wearing masks is 'shameful, not cool, a sign of weakness' (Capraro and Barcelo, 2020: 1). However, Bhasin et al. (2020) argue that gendered language is used on both sides of the mask debate. Mask wearers are often characterised as women blindly following orders, while anti-maskers are feminised as overly emotional in their irate and irrational reactions.

Internet debates and memes

Given the extent to which COVID restrictions and physical distancing measures have resulted in more people working from home, and lockdown measures have kept more people indoors, it is unsurprising that social media have come to play a central role in the unfolding of these debates and tensions around masks. However, early research suggests traditional news media remain important in shaping debates and public opinion about COVID masks. A US study of local news media found an increase in anti-mask reporting and a decrease in pro-mask news stories in US states following the enactment of a mask mandate (Morrow and Compagni, 2020). Internet memes have also been important media for expressing people's affective responses to wearing a mask. Early in the pandemic, a range of memes conveyed the frustrations of attempting to source a mask: often by showing images of people using ludicrous substitutes, such as disposable nappies, plastic bags, gas masks or motorcycle helmets. The suggestion of such images was that people would try almost anything to protect themselves against contagion, but also perhaps making fun of the idea of mask wearing.

Both anti-mask movements and pro-mask advocacy have largely taken place online, with the largest movement, #Masks4All, mobilised on Twitter, its own website and across other platforms. The #Masks4All movement voices the urgency for widespread availability of masks and wearing of masks through their website and key advocates, as well as those who join the movement by posting on social media using the hashtag. On its website, #Masks4All (2020a) has collected data on the recommendations and regulations of face masks in public spaces globally. It has also provided a constantly updated interactive world map showing the countries which had some kind of national requirement for masks to be worn and in what situations they were required. The #Masks4All website states that '[c]loth masks can help stop the spread of COVID-19, save lives and restore jobs', and collates resources about mask mandates in the USA and worldwide, scientific evidence for wearing masks, alongside information on how to make and effectively clean a cloth mask (#Masks4All, 2020a). Writing for *The Guardian*, one of the pioneers of the movement, entrepreneur Jeremy Howard (2020) attempted to urgently convey to readers a darker version of the 'we're all in this together' narrative: 'Covid-19 moves like a silent assassin, with unwitting accomplices. Maybe you'll be one of them. The best way to ensure that you're not: wear a mask and keep your distance from others'.

Sovereign citizens and similar movements are organised through material distributed online and on social media (Berger, 2016). In May 2020, a Singaporean woman was arrested for refusing to wear a face mask and was captured on video shouting 'I am a Sovereign... It means I have nothing to do with the police. It means I have no contract with the police and they have no say over me' (Koay, 2020). A similar clash between a sovereign citizen over mandatory mask-wearing was captured on video in Australia, when a woman filmed her own encounter arguing with employees at one of the outlets of the well-known Bunnings chain of hardware stores after being asked to wear a mask (Gillespie, 2020). She stated that she refused to wear a mask because it was her right as a 'living woman' to not wear one. The widely shared video was followed by a flood of memes as Australians took to social media to mock the woman dubbed 'Bunnings Karen'. This title is a reference to the popular Karen meme, referencing a privileged white woman who is given to harassing people in customer service and people of colour. 'Bunnings Karen' memes mocked the woman's implication that wearing a mask was a breach of her human rights and referenced the fact she was debating these rights with frontline workers at the store, likely underpaid for such labour (Lenton et al., 2020).

Anti-mask protests are largely organised and observed through social media, including the use of videos of clashes with police and other citizens. Hashtags like #nomask were used to voice frustrations and objections to mandatory mask-wearing, and express scepticism about the existence of the novel coronavirus and COVID (Bhasin et al., 2020). Facebook groups have been established by people against 'the face mask cult and its followers', as enshrined in the title of one such group. They and other social media users make and share memes to circulate on the internet. The memes disseminate anti-mask messages emphasising the supposed loss of individuality, conformity, stupidity or government control that mask wearing and its adherents symbolise. These memes include images of sheep with masks, comparisons of mask wearing people with zombies, referring to masks as 'corona diapers' and portrayals of children wearing masks as 'brainwashed', subjected to 'an array of physical and psychological trauma' and 'conditioned for servitude'. Meanwhile, on Instagram, a group of 'mom influencers' have conducted their own campaign against COVID masks, with a particular emphasis on alleging that mask wearing will harm children's health as well as flout their sovereign citizenship. These influencers often support COVID conspiracy theories and have ties to anti-vaccination campaigns, adopting similar messaging concerning the importance of avoiding 'invasive' health prevention practices such as vaccination and mask wearing that are represented as 'unnatural'. Parents are exhorted to 'un-mask our kids' and 'let kids breathe'.

The politics of co-becoming with masks

Mask politics draw attention to the ways that these complex events cannot be attributed to a single actor but rather require us to look more carefully at the way that agency *emerges* from within relations of discourses, objects, bodies, habits, relations of power and affects. Indigenous and First Nations scholars have long critiqued Euro-Western humanism for conceptualising agency as something possessed by humans and exercised over others, a way of thinking that is crucial for the operation of colonial power and oppression through control of the category of human (Watts, 2013; TallBear, 2014; Bawaka Country et al., 2016; Todd, 2016; Hernández et al., 2020). Watts (2013: 21) contrasts this Euro-Western thinking with 'place-thought', drawn from Native American Haudenosaunee and Anishnaabe cosmologies, which 'is based upon the premise that land is alive and thinking and that humans and non-humans derive agency through the extensions of these thoughts'. Similarly, the concept of 'co-becoming' from Indigenous philosophy posits that 'everything exists in a state of emergence and relationality' (Bawaka Country et al., 2016: 456). This approach does not mean unlimited openness, but rather suggests there are structures and patterns in this emergence which allow us to make sense of more macro-level processes at the onto-ethico-epistemological level.

Karen Barad's (2007) more-than-human performative approach is also instructive here in providing language for reframing agency in events. As noted in Chapter 1, Barad proposes the term 'intra-action' as a way to think through the multiple, emergent agencies at play when an event occurs. As she explains, a term such as 'interaction' assumes there are separate and individual agencies that exist prior to their interaction with each other. The concept of intra-action challenges this assumption, arguing instead that these distinct agencies emerge from rather than precede the interaction. Barad (2007: 33) contends that 'agencies are only distinct in relation to their mutual entanglement; they don't exist as individual elements.' This perspective aligns with the concepts of Place-Thought and co-becoming in emphasising the conditions of possibility rather than only actors and their actions and offers a way to draw together seemingly a wide array of disparate forces, both human and nonhuman.

The events traced in this chapter surrounding face masks during the pandemic can be productively understood through the concept of intra-action, as it allows us to conceptualise the agencies at play in their entanglement in the COVID crisis. Taking the viral video of 'Bunnings Karen' as an example, we can see the importance of Place-Thought (Watts, 2013): that is, starting with relations rather seeking to locate the meaning of the event in structural forces *or* individual actors. Bunnings Karen as an event, and a widely shared video, emerges from the intra-actions generated with and through discourses of individual freedom, state control, conspiracy theories around the novel coronavirus and the affordances of digital devices and social media. Many viewers of the video, however, would not have had any knowledge of the sovereign citizen movement at the time it was circulated on the internet and social media. We must also consider the collective affective environment of the COVID crisis as part of the conditions from which this event emerges, as well as the agencies of the novel coronavirus itself. These intra-actions take place in an environment in which the COVID crisis had heightened awareness of our collective vulnerability to each other's more-than-human bodies: in other words, our 'embodied sociality' (Bracken, 2020). Such conditions bring forth more-than-human assemblages constituted through constantly shifting boundaries, their breath and other bodily fluids extending the 'risky' space they take up.

This event is also is made possible through the capacities of the smartphone camera the woman used to film her encounter, as well as the well-known location in which she made the video. The shaky video is our only glimpse of the woman later dubbed 'Bunnings Karen', who is not visible until later filmed by another bystander when she is arrested. Social media affordances that facilitate loosely affiliated communities forming around terms like 'sovereign citizen' and the figure of the complaining white woman 'Karen' stereotype are parts of the intra-actions of these assemblages. The rapid sharing of videos on social media intra-acted with the Australian mainstream news media to circulate the video further. Another intra-acting agent within this event is the recently implemented mask mandates in place in the Australian state of Victoria where the video was filmed. As a result, the video shows a hardware store in which all visible staff and customers are masked. We should also consider the setting of the specific chain of Bunnings hardware stores: described by some people tweeting about the event as an 'Australian cultural institution' (Lenton et al., 2020). The event is further contextualised by Australia's colonial settler history, in which the protagonist in the video, as a white woman, would feel emboldened by a white supremacist society in ways in which Aboriginal people and other people of colour in Australia would not.

Examining these intra-acting elements of the 'Bunnings Karen' video can start to provide a more complex picture of how a moment in Australia's COVID crisis timeline involves a relational entanglement of the material objects, government policies, places and spaces, memes, digital devices and social media assembling to generate affective forces and agential capacities as part of COVID mask politics. In this more-than-human assemblage, a complex admixture of people, space, place, digital media and devices, COVID risk, health policies and mask politics come together in particularly vibrant and forceful ways.

Conclusion

Tracing the emergence of key debates around the meaning of the COVID face mask, we can see that there are significant macro-political issues surrounding health organisations and expert advisors, mask availability, production and supply chains, surveillance technologies, government policies and political groups on both sides of the debate around the benefits and problems with mandatory mask wearing. At the same time, micro-political events such as protests and viral videos of anti-maskers or coronavirus Karens reveal the ways these broader issues play out in everyday settings. We have argued for a more-than-human approach to understanding the politics of masks, using Indigenous and First Nations cosmologies to trace the diverse forces that produce the complex and contradictory meanings of masks during the first year of the COVID pandemic. The political events we have traced here around COVID masks emerge from and with a rich and vital assemblage of humans, places, technologies, platforms, ideologies, practices, affective forces, relational connections and knowledges that together generate or close down capacities for action. It can be difficult in such an overwhelming and shifting mess of events to see how they can be interrelated without immediately seeking to resolve them by way of an overarching narrative or explanation. Indeed, in many

ways, the COVID crisis, with the constant uncertainties and shifts in knowledges that have become our everyday experience, forces us to draw on ways
of thinking that embrace this messiness and complexity.

3 Living with Face Masks

Introduction

Face masks and their contested but central role during the COVID crisis offer us an opportunity to consider our intimacy with ordinary objects. How does an object transition from being new and unusual for most of the community to then becoming an ordinary thing? We have observed the COVID face mask making this transition in many communities during the crisis: moving from a specialised medical object to becoming a symbol of the continuing pandemic event and, eventually, an everyday object that many people, by necessity, became accustomed to wearing or carrying throughout their day. Part of this adjustment involves the development of new social norms, habits and routines that involve the COVID mask. This process of adjustment can be an uncomfortable one, in working towards achieving the kind of 'new normal' that has often been used to describe our post-COVID world. This chapter will explore the way this intimacy, as an embodied and affective experience, is connected to the context of COVID mask wear. Drawing on domestication scholarship, we contextualise face masks within a history of intimate objects that have become 'domesticated'. We then explore in depth the process by which COVID masks have become mundane objects, tracing shifting social norms, developing daily habits and unintended uses. We then consider how domestication theory might be extended by a more-than-human theoretical approach in answering the question of how COVID masks co-become with human bodies.

COVID masks entering everyday worlds

In this post-COVID era, mask wearing recommendations and mandates around the world have made mask wearing a new part of life for many people. In some countries (even those where there was no previous culture of mass masking), mask wearing has increased dramatically. In Australia, for example, in the early months of the COVID crisis, 17% of adults surveyed reported wearing a mask regularly. By September 2020, this number had jumped to 66% of respondents who said they had worn a mask in the past week. In Victoria, the only state where mask wearing in public had become mandated due to a second outbreak, almost all respondents (97%) reported wearing a mask in the past week: a clear example of behaviour change due to the introduction of mask wearing legislation (Australian Bureau of Statistics, 2020).

https://doi.org/10.1515/9783110723717-004

In addition to public health messages about the importance of mask wearing, popular culture now includes a range of representations of people wearing masks as they go through their usual daily routines: from news reporting to television drama. Whether we ourselves regularly wear masks or not, we can often see others around us with masks on, conducting their everyday activities (for example, the people exercising in Sydney in Figure 3.1). The 2020 festive season also brought with it a range of COVID-themed Christmas decorations. These included tree ornaments with mask-wearing Santa Clauses and reindeer or messages proclaiming 'We survived 2020', accompanied by the COVID symbols of toilet paper and face masks, as well as jolly face masks bearing well wishes or Christmas-themed patterns.

Adopting mask wearing as a safe and effective routine habit is not a simple process. The practice and the sensory experience of wearing a mask requires some getting used to. Given the diversity of the size and shapes of people's faces, not all masks fit all people well. People need to source and experiment with different designs and styles to find a mask that sits properly and comfortably on their faces. For some people, learning how to hand-craft a cloth mask is a priority (see more on this in Chapter 5). There are many complicated rules and guidelines for COVID mask wearing that must be learnt and routinely followed: for example, those provided by health authorities such as the WHO (2020h). They include selecting the most effective and best-fitting type of mask but also one that feels comfortable to wear, as well as learning how to don and doff masks safely to avoid cross-contamination, how often to change a mask during extended periods of wear and how best to store soiled masks

Figure 3.1: People wearing masks while exercising in Sydney. Photo credit: Katie Trifo, Unsplash.

and launder reusable masks. As we discuss further in Chapter 4, wearing a mask can affect how we breathe; and in Chapter 6 we argue that there are important issues to consider in terms of the environmental impact of COVID masks.

In Australia, when the Victorian state government mandated masks during that state's second outbreak, their guidelines stated that '[y]ou must carry a face mask with you when leaving home, even if you don't need to wear it while undertaking your current activity' (Department of Health and Human Services Victoria, 2020). Masks suddenly were made to 'live' alongside other mundane objects such as house keys and wallets: items which have established themselves as essential to everyday life over many years rather than days. These essential objects are part of the 'inner ring' of things that we hold close to our bodies for at least part of the day, which also includes objects such as clothing, shoes, jewellery and eyewear such as sunglasses and prescription spectacles (Agar, 2013). More recently, this inner ring has also come to include digital objects such as smartphones, travel cards and smartwatches: things that sit alongside our skin or in our pockets or bags to the point that we hardly notice their presence – or indeed, can feel their absence acutely (Lupton, 2015). As Agar (2013: 9) suggests, '[y]ou can tell what a culture values by what it has in its bags and pockets'. We do not constantly feel our clothing against our skin or our smartphone in our pocket; instead, these sensations become part of taken-for-granted embodiment. As the face mask becomes an object we must always carry with us, we can begin to interrogate precisely what new practices this newly valued object is embedded within.

Given their closeness to us, we could describe our relationship with these 'inner ring' objects that we carry with us as intimate dimensions of our selves as well as our bodies. In the social sciences, although there has been debate around the term, what we think of as intimacy has often focused on 'close personal relationships' with other humans. Closeness can refer both to closeness in interpersonal relationships as well as physical closeness, bodily contact or proximity (Jamieson, 2011). However, taking this notion of intimacy further, some scholars have sought to reimagine intimacy beyond the personal, conceptualising this bodily entanglement as extending to objects and nonhumans (Southerton and Bruce, 2019). It is this kind of intimacy – the ways that objects and bodies are entangled – in which we are primarily interested in engaging for understanding the sociomaterial dimensions of face masks.

Domesticating COVID masks

The process by which an object becomes part of the inner ring, or indeed even part of ordinary household use, has been described by scholars as 'domestication'. Scholars such as Roger Silverstone, Eric Hirsch and David Morley (2005) established the foundational concepts of domestication theory, arguing that understanding *how* technologies come to be part of domestic life is crucial to understanding their embeddedness in social practices. Domestication theories thus highlight the newness and adaptation of technologies, as well as their becoming-mundane: a continuing process that is never complete (Berker et al., 2006). Silverstone and Leslie Haddon (1996) emphasise that although the domestication approach takes the home as its initial focus, it has broader implications for thinking about how objects become familiar and adapted within daily life.

As we noted in Chapter 1, domestication scholars argue that the process by which an object becomes domesticated involves four elements: appropriation, objectification, incorporation and conversion (Silverstone et al., 2005). Face masks became *appropriated* from specialist medical technologies into a personal medical device through the event of the COVID crisis and changing public health advice about the value of mass masking. As we will discuss in Chapter 5, the widespread availability of cloth masks through major retailers, local makers and online platforms like Etsy, as well as the emergence of a DIY culture of mask-making, have contributed to rendering cloth face masks a household object. *Objectification* involves the location and arrangement of the object within the home, with the concept of 'home' being somewhat extended with intimate objects given that they travel on the users. In the case of face masks, we can consider the places they may be located for convenience, whether it be in a pocket, near the front door ready to take before leaving the house, in a bag or in the car. A recent article in UK *Cosmopolitan* magazine, instructing readers to carry a spare mask for rainy days due to the health risks of wearing a damp mask, offers an example of the complexities of mask wear and placement (Savin, 2020):

> By now, grabbing a protective face mask to prevent the spread of COVID-19 along with your keys, phone and purse as you're leaving the house is (almost) second nature. But new guidance published by the World Health Organisation (WHO) has shared that when it comes to rainy days, we all need to be packing a spare face covering too...

The domestication of face masks is inextricably entangled with health advice and information, as this article demonstrates. Another article, this time in the *Huffington Post*, advised readers on where to store their masks in their home. The author warned that many of the commonly described habits of hanging masks by the front door, putting them in pockets or handbags were not compliant with ideal infection control practices (Borresen, 2020). With the COVID mask so deeply embedded within the risk of novel coronavirus transmission, objectification practices form within a context in which mask wearers consume significant amounts of public health messaging that shape their behaviour.

Incorporation refers to all the ways that the object is integrated into the user's daily life through frequent use, including both intended and intended uses, as well as the different capacities these uses provide for the user. The most significant capacity offered by face masks for wearers is, of course, a reduction in the risk of either contracting or passing on the novel coronavirus. However, this capacity also has a range of flow-on capacities depending on the specific context of the wearer. As discussed in Chapter 2, in many countries, mask mandates allow for movement out of the home only for those people who wear masks or those who can provide reasonable evidence that they are exempted. Other countries, or regions within countries, have implemented mask mandates in situations where physical distancing cannot reasonably be implemented, such as on public transport or in indoor settings. In the USA, for example, many retailers, including some of the largest national chains, have rules requiring customers to wear masks upon entry to their stores (Friedman, 2020). As such, mask wearing affords wearers greater freedom of movement and access to a wider range of activities than may be available to them when they are restricted to the home or places where they can keep an appropriate distance from others to avoid infection.

There are many situations, however, where mask wearing becomes difficult to incorporate into everyday routines: for example, the consumption of food and drink in cafes or restaurants, or entering a bank, where security rules enforce the removal of a face mask for identification purposes. In these contexts, people face a clash of expectations and norms about what is the 'right' thing to do. Alongside practices that have developed in order to wear masks for their intended purpose (reduction of viral transmission risk) there are many mask-wearing practices and uses of masks beyond the scope of their intended use. The incorrect use of COVID masks has drawn significant attention during the pandemic. Mask-wearing practices that greatly reduce the effectiveness of face masks, such as wearing a mask below the nose, or wearing a mask around the nose but leaving the mouth exposed, have been identified by health authorities as 'mask don'ts' (World Health Organization, 2020g). Collective frustrations with these 'mask-fails', 'mask-holes' and 'covidiots' have been vented on social media, with internet memes comparing poor mask wearing with leaving your genitals exposed by your underwear and highly-circulated news stories about face masks being sold that have a hole in them to fit a drinking straw (Sparks, 2020).

Finally, *conversion* refers to how the users of the object integrate the object into their sense of self and use the object to express themselves (Silverstone et al., 2005). As we argued in Chapter 2, masks have been used to communicate political messages or affiliation, or even objection to mask mandates through displaying anti-mask slogans. We can also see conversion in action in the vast array of designs of cloth face masks available for purchase, with different fashion choices available so the wearer can not only integrate their tastes into their mask-wearing practices but also publicly proclaim them. This may include selecting masks that show the wearer to be a supporter of a certain sports team or a fan of Marvell, Disney, Hello Kitty, Harry Potter, Jurassic Park, Sherlock, Dr Who, Star Wars or Lord of the Rings ... and many more. Some designers have incorporated COVID masks into high fashion: for example, as shown in Figure 3.1. An article on US fashion news website *Refinery29*, offering a list of 'fashion-forward' masks, describes the importance of being able to express personal style while remaining covered up:

> Masks may still feel like an unusual addition to our daily routines because of the way they conceal the parts of ourselves that are normally exposed, but that doesn't mean our identities have to be hidden too. By opting for the right print to suit your own aesthetic and making a purchase you know will benefit those in need, wearing a face mask can become more than just our collective responsibility – it can be an entirely new form of self-expression. (Randone and Spencer, 2020)

As this quote illustrates, the self-expression of mask wearing is connected not only to the aesthetics of masks but also to the sense of mask-wearing as a moral act in service of the community (see Chapter 6 for a more detailed discussion of these aspects). Just as the absence of a mask on a person's face denotes a set of meanings (such as lack of concern about personal risk of infection or others' wellbeing, or anti-mask ideology), the purchase and wear of certain kinds of masks becomes a way of performing selfhood. The mask purchasers can feel good about their fashion choice as an expression not only of their personal taste but of their commitment to protecting vulnerable others from viral infection.

Embodiments of COVID mask use

Domestication theory has largely devoted attention towards the meanings, norms and practices associated with ordinary objects. More-than-human theory can help rethink the relationship between our habits with these intimate objects. It is important to understand how we learn to feel comfortable and uncomfort-

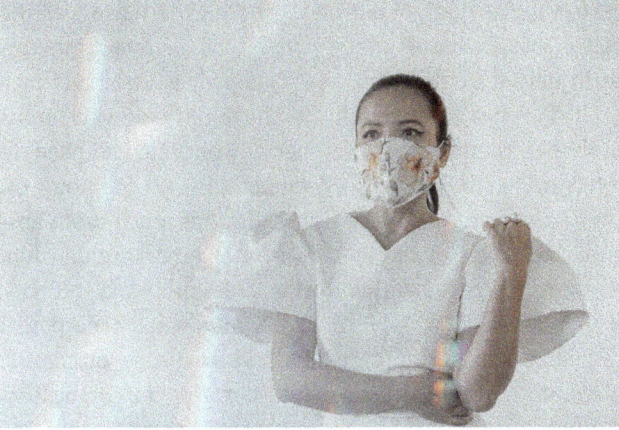

Figure 3.2: A mask design to complement a high fashion outfit. Photo credit: Andreas Weiss, Unsplash.

able with COVID masks through our bodily habits and capacities as we move through space and time. In the context of a pandemic, where masks have been implemented to reduce the risk of disease transmission through our breath, it seems more important than ever to confront how our bodies are constituted through our environments, rather than distinct from them. The theoretical shift away from privileging human agency in more-than-human scholarship has significant implications for how we conceptualise habits. Habits have often been conceptualised somewhat ungenerously as mindlessness or 'the disease of repetition that threatens the freshness of thought' (Malabou, 2008: vii). However, such conceptualisations do not make sense when we consider the humans and nonhumans and their agencies emerging from their interactions, rather than the human subject solely possessing agency. It is here that a focus on the distributed and relational nature of embodied practices such as mask wearing can be productive. Identifying the ways that habit sediments provides a way for understanding how masks become ordinary in both their meaning as a social object, but also their multisensory intra-actions with human bodies.

Braidotti's conceptualisation of the nomadic human subject recognises flows of affective intensities into and out from the enfleshed body: 'a folding-in of external influences and a simultaneous unfolding-outwards of affects' (Braidotti, 2008: 32). Braidotti (2008) argues that people's bodies warn them when limits or thresholds are reached as they are undergoing processes of transformation in response to other people or to nonhumans: they feel sensations or affective intensities that signal that something is wrong.

From this perspective, we can think about how people experience wearing COVID masks (themselves or observing other people) as feeling 'right' or 'wrong' (or somewhere in between). For President Trump, it is clear, the face mask is felt as 'wrong'. It does not accord with the image of healthy, vigorous masculine embodiment that he wants to project: especially once he had been diagnosed with COVID and admitted to hospital for treatment. His act of ripping off his mask as he stood on a balcony of the White House after his return from hospital was an unambiguous rejection of the idea that he or others around him needed protection from the novel coronavirus. However, as discussed in Chapters 3 and 4, wearing a face mask may (also) feel 'wrong' at the corporeal level even if people are committed to the principle of wearing one: prompting sensory feelings of discomfort, closing down capacities for communication with other people and altering embodied, taken-for-granted ways of being and relating in the world. Therefore the experience of wearing a face mask as 'right' or 'wrong' and the cultural meanings associated with masks and mask-wearing emerge from the entanglement of both material and social forces (Barad, 2007).

Faces are intensely communicative parts of the body, viewed as the most expressive parts of people's bodies and selves, with facial muscles, skin and organs all operating together to present to the onlooker a person's thoughts and feelings. Given that masks are worn very close to the body and obscure part of the face, the habits of ordinary social interaction must be renegotiated. With the lower half of the face obscured, modes of expression and communication are dramatically altered. Voices are muffled by fabric and a person's repertoire of facial expressions is dramatically diminished. Not being able to see people's faces can be affectively and relationally disorienting. French philosopher Emmanuel Levinas (1979) wrote about the importance of the face in our social relations. He argued that the face of the other 'invites me to a relation' (Levinas, 1979: 198), thereby presenting an ethical obligation to this relationship. Wearing a face mask disrupts this intimate face-to-face encounter and we are left to navigate social relationships, and our obligations to them, in new ways.

These major transformation in habits of sociality and embodiment and the intense affective forces that accompany them were vividly encapsulated in the words of US-based portrait artist and writer Riva Lehrer. Lehrer lives with spina bifida. Her work focuses on representations of people who live with marginalised and stigmatised embodiment, seeking to demonstrate their individuality and rendering them more visible in a society that tends to ignore and shun them. During the COVID crisis, Lehrer penned an opinion piece for *The New York Times*, writing that face masks disrupt her long-established practices of representing her subjects as well as her inspiration for creating her portraits. The piece was titled 'The virus has stolen your face from me'. For Lehrer, face masks are disguises that cut off access to more than half of a person's face, and therefore, access to much of their unique identity, biography and affective response to the world: essentially, their humanity. She writes that, as a portrait artist, she has a 'face hunger' and that 'Faces are my whole life. I think of the human face as a theater that performs the actor inside'. For Lehrer, therefore, as a person living with a disability rendering her more vulnerable to COVID and an artist, 'masks are both saving my life and ruining it' (Lehrer, 2020).

With facial expressions involving the mouth much more difficult to interpret while wearing a mask, the eyes become important for expressing and communicating with others. This has led to a renewed interest in the phenomenon of 'smizing', a term meaning 'smiling with the eyes' that was purportedly coined by former supermodel and popular culture figure Tyra Banks on her television show *America's Next Top Model* (Singh-Kurtz, 2020). Smizing is also related to the concept of the Duchenne smile, sometimes called a 'felt' smile, which is the expression of genuine enjoyment that not only lifts the corners of the mouth but transforms the shape and expression of the eyes. The Duchenne smile is not merely a perceived twinkle in the eyes but a muscular performance that occurs when the '[p]eriocular muscle pulls the outer corners of the eyebrows slightly downwards, produces bagging below the eyes and forms wrinkles to the corners of the eyes' (Surakka and Hietanen, 1998: 25). Importantly, facial expressions are not universal: research has suggested people from different cultures interpret and display facial expressions in unique ways (Mai et al., 2011). Considering these differences illuminates the importance of social, cultural, geographical and material specificities when we consider the meanings and experiences of COVID mask wearing.

'Maskne', the nickname given to acne caused by wearing masks, reminds us that adapting to mask wearing is a deeply embodied process whereby the COVID mask's unfamiliar intimacy with our skin, and indeed the porous nature of the boundaries of the body, causes a reaction over time through habits. Dermatologists have observed the phenomenon, suggesting that maskne may arise from the sweating that occurs on the face underneath a mask or the high humidity inside the mask, as well as inadequate mask washing that fails to remove bacteria that have accumulated during extended wear (Han et al., 2020).

Magazine and newspaper articles tout the latest treatments for maskne. An article in *Marie Claire* recently even highlighted a face mask (skin treatment) to treat maskne that can be worn underneath your mask (Freund, 2020). *The New York Times* published a guide to maskne, three tips on how to avoid it and what to do if you get it, with opinions from multiple dermatologists (Rubin, 2020). *British Vogue* has provided their readers with a list of 'Face Masks That Don't Cause

"Maskne"', including masks made from silk, bamboo or treated with anti-bacterial agents (Coates, 2020). Google Trends data show a huge surge in interest in the term 'maskne' towards the end of May 2020. It has been a popular topic on social media platform TikTok, where by mid-November 2020 the hashtag had received 46.5 million views. Many of the videos offer skincare advice and routines, with skincare already being a major content trend on across many platforms, while other videos are advertisements for maskne-combatting face masks. Other TikTok creators simply share their frustrations with their maskne, revealing their acne as they remove their face masks.

Maskne emerges as the tangible expression of human and more-than-human entanglements, of the generative intra-activity between fabric, flesh, and air. Its appearance draws attention to the more-than-human ways that people's habits are adjusting to mask-wearing practices. We can see shifting capacities of human facial skin to cope with the changing conditions, excreting excess oil in the increased humidity within the mask. There are also the agencies of the bacteria within the mask at play here, enlivened by the warm environment. Far from mindless or thoughtless, these habits draw attention to the limitations of locating thinking in the mind.

As masks are incorporated into our bodies, they also can reduce some bodily capacities that previously felt so effortless they barely warranted a thought for most people. The regular usage of face masks has drawn greater attention to the ways they become uncomfortable over time, with a great number of adaptive design improvements, add-ons and technologies becoming popularised to deal with these problems. Spectacle wearers, who have struggled with their lenses fogging up while wearing a mask, have found a range of DIY solutions to this problem, including using tape on the bridge of their noses or putting shaving foam on the lenses, and then cleaning it off to leave a protective film (Lockwood and Jordan, 2020). There are also plenty of masks available for purchase with nose wires included, to make glasses wearing easier. 'Mask brackets' are also being sold as an add-in device for your mask. These are devices usually made of plastic coated with silicone that sits between the wearer's face and the mask to create more space to breathe (Brickell, 2020).

Depending on the material used and layers of thickness, some face masks can make breathing much harder (see Chapter 4 for more discussion of this). In his autoethnographic account, US-based communications scholar Douglas Kelley (2020: 117) describes the sensations of wearing his mask for seven months after he had undergone a stem cell transplant prior to the pandemic, rendering him highly vulnerable to potential serious infection with COVID. In his account, familiar and deeply established bodily habits jar with new ones, as he struggles to incorporate the mask into his everyday bodily practices. Kelley vividly recounts feeling constricted and contained by the mask on his face, comparing the mask to shoes:

> The thickness of the mask can at times feel constricting, especially when I am exerting my body. This has meant accepting that it will often be harder to breathe through the mask when active, and I've learned to carry lip balm because of the dehydrating effect of each breath when wearing the mask for a prolonged period of time. The mask can also feel constricting in the same sense that shoes do for some of us. Personally, I can't wait to kick off my shoes when I get home from work – free and toe wriggling fresh! In the same way, once away from others, taking off the mask feels fresh and free, more fully engaging of the senses.

The sensory experience of communicating and speaking is also significantly affected by wearing a mask. Although there may be difficulties literally being heard while wearing a mask, as Mickey Vallee (2020) describes, the sensations of mask-wearing involve an auditory distortion that can be disorienting as one hears oneself. Vallee explains:

> The timbre of my voice is quite muffled, like I'm doubling up on the sound of my voice from the inside. Customarily, we hear two versions of our own voices: one from within the vocal cords that resonates in the skull, and one that is the sound of the voice as it leaves the mouth and resonates off of surfaces and in the air around us. With the mask, the inner voice and external voice are more closely intertwined. The voice thus becomes more internal. To be heard, I speak clearly, loudly; I enunciate, and I'm convinced by the seemingly seamless interactions I have that I am, in fact, succeeding at communication. (Vallee, 2020: 266)

Vallee highlights the way that voice is produced through an assemblage of parts of the body (the skull, the vocal cords, tongue and so on), with the air and objects in the vicinity. As he emphasises, the voice is an achievement of spatiality, not simply the individual's body. Face masks become part of this environment and the capacity for sound shifts, creating different resonances and sensations.

Considering alternative or non-normative bodily habits and capacities is also productive for thinking through why some people cannot wear face masks. Such people risk public harassment and judgement (especially when their disability or distress is not 'visible') and refusal of service in essential areas such as supermarkets (Pendo et al., 2020). There are many reasons why mask wearing can be difficult or distressing. Masks may be experienced as physically uncomfortable, especially for people with pre-existing conditions such as breathing difficulties or who have mental health conditions that may make the sensory experience of having the mask on the face disturbing. People with autism spectrum disorders, post-traumatic stress disorder, severe anxiety or claustrophobia may be severely disturbed by the sensation of having their nose and mouth covered. People with disabilities or pre-existing conditions that impact their mobility may not be able to safely wear masks as they are unable to remove them on their own (Seale, 2020). Survivors of sexual violence can find mask wearing (on their own faces or the faces of others) to trigger distress: it is difficult for them to disclose the reason for avoiding mask wearing to strangers (The Survivors Trust, 2020).

Mask wearing can also significantly delimit the bodily and social capacities of people with hearing impairments who rely on lipreading in their everyday lives. A report by Helen Grote and Fizz Izagaren (2020), both D/deaf medical practitioners based in the UK, describes the lack of consideration for the deaf community amid the fervour to adopt masks on a global scale:

> The lack of support has been one of the hardest challenges we have faced at work during the pandemic. It leaves us and our D/deaf patients feeling isolated and ignored. Reading articles and tweets about the importance of masks, with no consideration of the impact on the D/deaf, leads us to conclude that policy makers and academics have forgotten about the importance of equality impact assessments in this area.

This report highlights the difficulty of sourcing transparent masks for use in hospital settings in the UK and emphasise the need for health workers to consider the needs of their D/deaf patients. Their experiences and those of others who cannot wear masks, complicate the humanist narrative of mask-wearing presented by organisations like #MasksforAll (Chapter 2) as a moral good that protects others. It highlights the need to think relationally about how these practices intra-act with existing inequalities and modes of embodiment that differ from the norm and are often not considered when guidelines and mandates about COVID masks are devised and publicised.

Conclusion

As we have demonstrated in this chapter, the domestication or 'making ordinary' of COVID masks can be understood as a process through which new social norms and meanings are developed to attach to masks through changing practices, leading to the configuration of habits. We have sought to expand this inquiry with a focus on bodily habits from a more-than-human perspective. Over time, face masks become more familiar and consequently perhaps more comfortable. Through trial and error, some (but not all) people can find masks that work for their bodily affordances, capacities and embodied socialities; nose wire prevents spectacles fogging up, slogans or fan-based imagery make people feel better about wearing them, some fabrics and elastics are softer than others, some are better suited to smaller or larger faces or avoiding an outbreak of maskne. Bodily habits form and reform to facilitate the face mask into existing rich routines of movements and inclinations such that it becomes more familiar to reach for the mask on the way out the door, slip it onto our faces and remove it safely. This is to say, we also must get comfortable with the sensations of objects, the bodily habits and capacities they afford, as well as the sociocultural meanings within which they are embedded: and indeed, these phenomena are deeply entangled. Although COVID masks' entrance into our lives was abrupt, the slower processes by which we come to know and become comfortable with the objects that are most intimately close with our bodies are beginning to emerge in our developing relationship with face masks. For some people, however, mask wearing can never become familiar or accepted, due to disability, distress or simply feeling as if the mask does not 'feel right' on their face.

4 Face Masks and Breath

Introduction

In this chapter, we examine the embodied and sociomaterial dimensions of breath and breathing in relation to face masks and mask wearing during the COVID crisis. As the pandemic became a pronounced presence in our media feeds and daily lives, early public and medical concern focused on how the novel coronavirus is transmitted. This concern is understandable: knowing how the virus spreads allows for the identification and implementation of preventive measures. Of central importance to these and more general discussions of COVID is the notion of breath. Not only does COVID disease potentially restrict people's capacity to breathe, but the coronavirus is also spread *through* breath. As a result, breathing, a physiological process so essential to life, became associated with risk. We trace breath not as a universal physiological process but rather as a more-than-human phenomenon. COVID masks provide a material boundary of sorts between the body and its unstable surrounds, designed to limit invisible encounters between breath and other breathing bodies in close proximity. Yet our lived experience of wearing face masks exceeds such utilitarian understandings. Breath comes to matter in new ways. We consider how wearing a mask illuminates the porous or 'leaky' boundaries of the body and the implications of this porosity for our relationships with others. We also examine the social, political collective dimensions of breath as surfaced through the Black Lives Matter protests during the pandemic.

Contemporary understandings of human breath

Human breath can be understood as the physiologic process of inhalation and exhalation: a complex choreography of musculature and chemical and mechanical processes that both delivers oxygen to our cells and expels the carbon dioxide created in the process (Cedar, 2018). While humans can survive days without water and perhaps weeks without food, we can survive only a few minutes without breathing (Allen, 2020). Breath is therefore a window of sorts between life and death, constantly filling the body and leaving the body in an ongoing cycle of vital repetition. Breath enacts 'the ephemeral materialization of air at the interface of body and world' (Oxley and Russell, 2020: 1). While the physiologic processes of inhalation and exhalation are intricate, the experience of breathing goes largely unnoticed, often evading our attention unless it is im-

https://doi.org/10.1515/9783110723717-005

paired in some way, perhaps through illness, physical exertion or obstruction. Therefore, breath, as conceptualised within western thought, has been most extensively studied through medical and clinical perspectives concerned with the physiological structures and processes involved in breathing or its impairment (Malpass et al., 2019).

Although undoubtedly crucial, such perspectives overlook the social, metaphysical, political and affective dimensions of breath (Górska, 2016). In response, social sciences and humanities scholars have sought to expand these dialogues through examinations of breath (and air) as it relates to gender, cosmology, geography, political ecology and more-than-human ontologies. This corpus of work illuminates the complexity, materiality, and multiplicity of breath as it is experienced and made meaning of across cultures, histories, space and place. Feminist scholarship has also examined breath, with the work of feminist philosopher Luce Irigaray perhaps the most well-known (Irigaray, 1999, 2002). Irigaray criticises western philosophy for being 'founded in the solid' (1999: 2) and for 'forgetting' the element of air. She argues that air is what makes all thinking possible: by overlooking it, we also obscure the body and sexual difference. Looking to Eastern traditions and knowledges, Irigaray (2002) works to redirect western philosophy from its enduring concern with thinking to a philosophy of air that acknowledges the role of the body and breath in processes of knowing and being. Although her work has been critiqued for its essentialism, heteronormativity and orientalism (Górska, 2016), it offers an intellectual springboard for further analysis. In the edited volume *Breathing with Luce Irigaray* (Skof and Holmes, 2013), contributing authors critically engage with Irigaray's writings to provide analyses of breath in relation to (human) embodiment as well as relations of nature, culture, spirituality, sexual difference and interculturality.

Karen Barad (2007: 139) uses the term 'phenomena' to refer to so-called 'objects' of study where objects are not defined, bounded entities but rather emergent relations of concern produced through 'the ontological inseparability/entanglement of intra-acting agencies'. This approach allows us to examine breath as always corporeal, social and political. We are also inspired by Stacy Alaimo's concept of trans-corporeality, which she defines as the enmeshment of 'all creatures, as embodied beings … with the dynamic, material world' (Alaimo, 2018: 435). If we consider the embodied phenomena of breath and mask-wearing as trans-corporeal, we are prompted to consider how such phenomena are entangled with 'biological, technological, economic, social, political and other systems, processes, and events' (Alaimo, 2018: 436). We therefore take up this concept to think through the multiple ways in which breath comes to matter in the time of COVID.

Interdisciplinary approaches to the study of breath have more recently been embraced to account for breath's multi-faceted complexity (Malpass et al., 2019). The Life of Breath Project, based in the UK, brings together scholars from across the arts and humanities as well as artists and other health professionals to examine the multiple dimensions of breath and breathlessness (Life of Breath, 2020). Through creative scholarly and public engagements, the work of this collective examines the more 'invisible' cultural dimensions of breath and breathlessness to elaborate clinical perspectives. These scholars' work spans projects that explore breath and its role in speech and voice coaching, yoga and other breath-based practices, singing and performance, as well as breath as the physiological expression of nervousness and anxiety. Rebecca Oxley and Andrew Russell (2020) also provide a detailed anthropological account of the multiple meanings of breath by drawing on thousands of years of history and offering examples from Indigenous, First Nations and ancient cultures, western and Eastern religions, ritual ceremonies and working practices.

The numerous multi-media outputs emerging from the Life of Breath project underline breath as a life force, implicated in both living and dying, modes of expression and the cultivation of relationships between humans, environments and nonhuman animals. Perhaps most relevant here is the work of Magdalena Górska, who extends feminist approaches to breath and takes what she calls an intersectional posthumanist approach. Her book *Breathing Matters* (2016) provides an extensive analysis of breath as a social, material, cultural and physiological performance. Górska achieves this through in-depth engagement with specific cases such as black lung disease, panic attacks, and phone sex work. Through these examples, she traces the ways struggles for breath and what she terms 'breathable lives' are always corporeal/material political practices. In this conception of breath, the body and embodiment are central, but are reconfigured through a more-than-human lens.

We take inspiration from these scholars in our examination of how breath as a sociomaterial phenomenon has come to matter in the context of the pandemic. Doing so allows us to consider breath, and the embodied experience of mask wearing, as at once social, political, affective and corporeal. This approach also prompts us to attend to the material and discursive specificities of breath as well as the sociopolitical contexts through which meanings, practices and experiences of breath emerge with and through the COVID mask.

Breath in the context of COVID

Breath has played a central role in both popular culture and biomedical discussions of the COVID crisis. Medical knowledge helps to explain how the novel coronavirus affects the body: once the virus enters the body, it often attaches to the lungs, which can prompt inflammation and cause the small air sacs that make up the lung (alveoli) to fill with fluid. This condition is known as pneumonia, which can cause difficulty breathing (Shi et al., 2020). The virus can also cause clotting in the small blood vessels of the lungs, which further impairs breath. Consequently, shortness of breath and coughing are two of the key symptoms of COVID (along with fever) for which members of the public are advised to watch and to seek testing if they experience these symptoms (World Health Organization, 2020h). In severe COVID cases, breathing can be so dramatically impaired that intubation and mechanical ventilation are required. Intubation involves placing a tube into a person's windpipe so far as to reach the lungs, then attaching the tube to a mechanical ventilator to keep oxygen flowing freely (Readfearn, 2020). This procedure is not unique to the treatment of COVID disease: it also used to treat severe cases of non-COVID-related pneumonia and illnesses such as chronic obstructive pulmonary disease. However, in the context of the pandemic, ventilation has become more prominent in the social imaginary and largely associated with severe illness from novel coronavirus infection.

The pathology of the novel coronavirus therefore implicates both breath and breathlessness, concepts that are increasingly woven through COVID-related conversations. In one example, concern over the virus's ability to impair breath manifested in a 'fake news' piece that circulated widely on social media in the early days of the pandemic. The post, which was shared more than 30,000 times in over a dozen countries, suggested that holding one's breath for ten seconds was an effective test for COVID-19 (Factcheck, 2020). Specifically, the author of the post claimed that if a person could breathe deeply into their lungs and hold it easily for ten seconds without coughing, they were unlikely to be infected with the coronavirus. Falsely claiming to be from Stanford University, the author also suggested drinking water regularly can help prevent the disease. The WHO and other expert sources explicitly denied these claims (World Health Organization, 2020i), yet the circulation of this content underscores the focus on breath and breathing/breathlessness within COVID related discourse and experiences during the pandemic.

Breath is also linked to the spread of the coronavirus. Transmission is said to occur through contact with respiratory droplets and aerosols which are expelled when an infected person coughs, sneezes, talks, laughs, sings or otherwise exhales heavily. This type of transmission can occur when a person is within one metre of someone who is infected. Larger droplets are heavier and fall to the ground quickly, while aerosols may linger and travel further (Centers for Disease Control and Prevention, 2020b). While the risk of droplet transmission has been widely acknowledged, the role of airborne transmission via aerosols has been more highly contentious. It was only in October 2020, more than six months after the pandemic was declared, that the CDC listed aerosols as a method of transmission on its website and suggested that airborne particles potentially carrying the novel coronavirus can remain suspended and travel beyond 1.5 metres. This stands in stark contrast to the CDC's earlier advice downplaying the role of aerosol transmission, which in turn holds implications for recommendations for preventive measures (including the importance of mask wearing to minimise viral spread). This evolving dialogue highlights the fluid and unstable meanings of breath as they emerge through the rapidly shifting sociopolitical and scientific knowledges, contexts and practices of the COVID crisis.

It is now largely agreed in public health organisations that face masks mitigate the risk posed by breath by acting as a physical barrier that catches the respiratory droplets expelled from the mouth or nose. This barrier prevents these droplets from travelling as far usual and from being either inhaled or landing in the mouth or nose of others nearby (Centers for Disease Control and Prevention, 2020c). Researchers have shown that triple layer surgical masks and cotton masks, including home-made masks when made from tightly woven cotton, are the most effective at reducing exposure to infectious droplets, while neck gaiters and folded bandanas are the least effective (Zangmeister et al., 2020). In these largely clinical discussions, effectiveness emerges and is performed through the intra-activity between multiple material agencies (breath, fabric, air) and discourses of risk and infection.

Despite the protective role of face masks, resistance to mask wearing has taken multiple forms and expressions (outlined in Chapter 2). While much of the rhetoric of this resistance has revolved around sovereign rights and freedoms, breath itself has been co-opted to create claims that mask wearing is risky. One example is the claim that wearing a mask will expose wearers to lower levels of oxygen and increased levels of carbon dioxide accumulated as they breathe out, and thus make them ill. This myth has been soundly debunked on reputable health websites such as Mayo Clinic and the WHO 'mythbusters' site (Mayo Clinic Health System, 2020; World Health Organization, 2020i). While these responses to the resistance of mask wearing draw on expert medical knowledges, other responses engage the affective capacities of social media memes in efforts to make a compelling point. One meme circulating on social media says, 'If you don't like wearing a mask, you're going to hate the ventilator': accompanied by an illustration or photograph of an ill person lying prone and

Figure 4.1: Breath becomes more visible in certain climactic conditions. Photo credit: Pavel Lozovikov, Unsplash.

intubated with a ventilator. In contrast to the public health approach which mobilises scientific knowledge, these memes seek to arouse strong affective forces – and in this case the macabre – to make a point.

New vocabularies and materialities of breath

Placing a mask on our face is deeply corporeal, sensory, and affective. It disrupts our usual ways of being in the world and reconfigures our relations with others. So too does it alter our experiences of breath: we may suddenly feel our breath as it condenses on our skin, we may be confronted with its smell, we may 'see' it as it fogs up our spectacles. Consequently, we may notice our bodies, our body boundaries and their 'leakiness' (Grosz, 1994; Shildrick, 1997) in new and uncomfortable ways. As scholars observed pre-COVID, breath is largely experienced and conceptualised as homogenous and invisible, and often evades our notice (Allen, 2020). This echoes Irigaray's (1999: 5) observation that 'air does not show itself. As such it escapes appearing as (a) being. It allows itself to be forgotten'. As we understand more about the transmission of the novel coronavirus and how it affects people, new vocabularies of breath and air emerge, with renewed attention paid to the materiality of breath and its capacities.

The tiny droplets in human breath are sometimes visible to human eye, as in cold conditions when exhalations condense in the air (Figure 4.1), but for the most part, we perceive our own breath and that of others using other sensory cues, such as touch, smell or sound. The invisibility of breath means the threat to our health and that of others posed by breath (or air) may be difficult to grasp for its intangibility. Pre-COVID, public health campaigns have often sought to materialise the risk of viral infection through breath. Many of us have seen iconic images of these aerosols sprayed in a large contagious cloud from a person's mouth or nose when they are coughing or sneezing. These images are frequently used in campaigns to illustrate the risk of being too close to an infected individual and the importance of personal hygienic measures such as wearing a face mask. They have been manipulated so that the tiny particles flying through the air are visible to the human eye. In real life, the risk is invisible: and all the more potent and frightening for it. One example is in a poster used to combat the spread of swine influenza by the UK Department of Health in 2009. It used a close-up photograph of a man (in one version) or a little girl (in the second version) coughing or sneezing out droplets and aerosols (coloured white). The poster text noted that 'When you cough or sneeze, your germs go everywhere. Fast.' Face masks were not mentioned as part of prevention in this 'Catch it. Bin it. Kill it.' Campaign. Instead, Britons were advised to use a tissue to 'catch' the germs when they cough or sneeze, to place the used tissue in a garbage bin as soon as possible and to then wash their hands with soap and water to kill any germs remaining.

The need for masks to mitigate this invisible risk posed by breath is not always obvious or resonant enough to compel action. In a culture that is, as Irigaray points out, preoccupied with logocentric ways of knowing, not being able to 'see' or observe the risk obscures its legitimacy. This may be reflected in part by the widespread resistance to mask wearing discussed in Chapters 1 and 2. This dilemma has been well noted by public health and health communication experts and a variety of infographics and video animations have thus emerged in efforts to make visible the role of breath and face masks in both spreading and protecting against the virus. One Australian government health campaign, entitled 'How the virus spreads', sought to warn asymptomatic young people about their role in spreading the novel coronavirus. The advertisement, released in August 2020, showed infected young people coughing or sneezing and transferring the virus by touching objects or food shared by others. The places touched by these people were rendered in a bright glowing pink colour in an attempt to make the viral risk more obvious. In October 2020, the *New York Times* published a video animation that provides an in-depth explanation and demonstration of how masks work on their website (Fleisher et al., 2020). The text and the visuals provided work together to clearly explain and illustrate the so-called mechanics of masks in fine detail. Similarly, *The Scientist* magazine (Kwon, 2020) has published a colourful infographic on its website illustrating what happens when breath encounters a face mask, using visual elements to show how breath circulates in the air and how it 'touches' and therefore infects other bodies if not protected by a mask.

These visual representations contribute to new knowledges and imaginaries of breath as they emerge and circulate through the sociomaterial contexts of COVID. In an essay appearing in *The Guardian*, author Jennifer Mills described her experience of the pandemic in Turin, Italy, and elaborates on the increasing visibility of breath that emerges through COVID conditions (Mills, 2020). She wrote:

> As I stepped into the middle of the road to keep my distance from someone, or stood a metre away from a person I was in conversation with, I realised that I saw my own breath, and the breaths of others, as clouds of vapour. I imagined airborne droplets around us, filled with some of the microbial life forms – bacteria, viruses – that lived in, and formed, our bodies. (Mills, 2020)

In Mills' account, breath is clearly recognised as an agential assemblage teeming with tiny life-forms with the capacity to touch and contaminate others. Breath leaves behind it trails of contagion that call into question the assumed boundaries of the body (Thorpe et al., 2021). As a result, our encounters and relations with the bodies of others are altered and increasingly performed through improvised choreographies of avoidance. Dance critic Gia Kourlas elaborates on this idea, noting how fear of contracting the novel coronavirus and of the exhalations of other bodies is expressed through bodily comportments, altering the way people engage with each other in public spaces. In a *New York Times* article entitled 'How we use our bodies to navigate a pandemic', Kourlas (2020) observes, 'Now the choreography of the streets has taken on higher stakes. It's the difference between health and sickness, life and death'.

The sensory and physical dimensions of breath are central to how COVID masks are experienced and made sense of. When we wear a mask our breath remains close, intra-acting with other material agencies to condense on the inside of the mask and form a moist residue on our skin. This experience can be particularly unpleasant in hot weather. An article appearing in the *Chicago Tribune* identifies this as the 'yuck factor' and columnist Nina Metz summarises, 'It's gross. It's hot. It's wet' (Metz, 2020). The so called yuck factor or 'grossness' of breath in the context of COVID exceeds discursive understandings of breath as contagious or infections. Rather, this 'grossness' is deeply material, resulting from the transformation of breath as an invisible vapour that usually floats away from the body to something that can be smelled, seen, and felt. Extremely cold weather, conversely, can lead to the moisture from breath to freeze on a mask, rendering it more difficult to breathe through. Here again, the broader spatial and climatic conditions in which COVID masks are worn are integral to how they feel on our faces and how well we can breathe while wearing them.

When we notice our breath as something that can be seen and felt – or in other words, when the 'yuck' factor of breath reveals itself – we may be confronted with the materiality and porousness of our bodies in new ways. When we notice our bodies as 'leaky' or porous, the Cartesian assumption of the body as a discretely bounded entity is raised for further examination. Feminist scholars have explored this so called 'leakiness' of the body in pointed critiques of the medical model and its inherent embracement of Cartesian dualism (Grosz, 1994; Shildrick, 1997). Modern medicine mobilises this thinking to further to categorise bodies according to strict dualisms (healthy/sick, normal/abnormal, male/female). Within this framework, bodies that are viewed as departing from idealised tightly contained, controlled body are imagined as unruly, 'leaky' and with unstable boundaries that disrupt the order and health of society (Shildrick, 1997; Lupton, 2012). Yet in the COVID crisis, the leakiness of *all* bodies becomes apparent; we all breathe, we all speak, we all cough. Breath emerges as 'a leaking, uncontrollable, seeping liquid ... a disorder that threatens all order' (Grosz, 1994: 203). Face masks are thus introduced to preserve this order, and therefore human life, by reinforcing these bodily boundaries and containing bodily exhalations.

Alaimo's (2016, 2018) concept of trans-corporeality also contributes to our thinking about breath, the body and its exchanges with other human and non-human entities. Trans-corporeality challenges the Cartesian notion that clear boundaries exist between bodily interiors and exteriors. The problem with this assumption, Alaimo suggests, is that we come to understand ourselves as completely distinct from other people and other living and non-living agents surrounding us (or living within our bodies, such as viruses and bacteria). As a result, we overlook the many ways we affect and are affected by human and non-human others and the world around us. A trans-corporeal account of breath unsettles these problematic distinctions as it acknowledges breath as a vital interface between bodily insides and outsides, between self and other. When we feel our breath as a sweaty film on our face or as ice stiffening a mask or as we shrink away from other breathing bodies on the sidewalk, we enact an acknowledgment of our porous bodily boundaries. When we inhale, we draw external matter in, and when we exhale, matter from our bodily interiors is expelled into external atmospheres. Both our bodies and our environments are touched and transformed in the process, bringing the shared collectivity of our social and embodied lives more clearly into view.

'I Can't Breathe': the politics of breath

In the southern hemisphere summer preceding the outbreak of the COVID pandemic, Australia experienced extensive bushfires that caused widespread devastation. It was a summer marked by stifling smoke and peculiar orange-tinged skies. Face masks of the P2 specification (also known as N95 masks), air purifiers and household fans sold out quickly across the country as people sought to breathe well when the air outside continually reached hazardous levels for the better part of two months. As is the case in the COVID crisis, breath was of central concern. Public health warnings regularly cautioned those with asthma and other respiratory conditions to stay indoors with windows sealed. In addition to the startling and eerie atmospheric hues caused by bushfire smoke, the sight of people wearing face masks became more commonplace in major Australian cities as people struggled to carry out daily routines safely. Wildfires also ravaged parts of California later in 2020, with similar effects on the air quality in those regions. In her article reflecting on the entanglements of these fires with the pandemic, Mel Chen (2020) notes, 'COVID-19, and the California seasonal fires, intensified due to human-induced climate change as well as poor fire policy, are two airy phenomena that dance together, aloft, touching'.

Air, like breath, is not a homogeneous entity. Rather, air is metaphysical, ecological, and central to the sociopolitical and economic conditions of daily life. Breath therefore emerges as part of an emerging contemporary philosophy and cultural politics of air concerned with air as a space of warfare, a site of pollution, and as implicated in and affected by processes of climate change (Braidotti, 2020). While we all breathe, we do not all breathe the same air. Many places in the world are exposed to high levels of air pollution and for those living there, face masks are already part of their everyday routines (Flaskerud, 2020; Ma and Zhan, 2020; Horii, 2014). As Chen (2020) observes, during the wildfires in California as they raged through the pandemic, air overlaps with the COVID pandemic in complex ways. There, as in Australia during the bushfire crisis of the 'Black Summer' of 2019–2020, P2 and N95 face masks were being rapidly bought up by the public, diminishing the supplies for healthcare and other frontline workers.

Both breath and air, therefore, are deeply material *and* political: those who can breathe clean air and how freely they can breathe it is unevenly produced through intersecting social, geographical and political forces. We do not all breathe in the same way. We do not breathe the same air, nor are all bodies afforded the same opportunities to breathe. This point was starkly illustrated during the pandemic by the murder of Black American man George Floyd in a Minneapolis street in May 2020. While arresting Floyd, police officer Derek Chauvin forcibly restrained him, pinning him to the ground and pressing his knee into Floyd's neck for eight minutes and 46 seconds. During this time, Floyd uttered the words, 'I can't breathe' eleven times until he was rendered unconscious and then expired from lack of oxygen. Images of Floyd's death circulating in the news media and on social media generated protest marches and rallies in the USA and elsewhere as part of the Black Lives Matter movement in May-June 2020. His dying words offer a haunting refrain echoing those of another Black man, Erin Garner, who died in 2014 at the hands of a New York City police officer, who put him in a chokehold while arresting him. The same words have been uttered by many other Black, Indigenous and other people of colour who have experienced violence at the hands of white authority, including David Dungay Jr, an Indigenous Australian man who died in custody in a Sydney prison days before his release in 2015. The story of Dungay's death is harrowing, the transcription of his final words difficult to read. They conjure a painful imagining of a young man pleading for the ability to live. 'I can't breathe. Please, I can't breathe,' Dungay repeated over twelve times in the last nine minutes of his life. His pleas were repeatedly ignored by multiple white male prison guards, resulting in his death (Davidson, 2020b).

In response to Floyd's murder and the pervasive violence against people of colour communities, the Black Lives Matter mass protests erupted in the USA and in many countries around the world and continued for weeks, infused with a renewed and palpable urgency. Given the pandemic conditions, protestors were encouraged to wear masks, and most complied, some donning face masks with the words 'I Can't Breathe' emblazoned across them (as shown in Figure 4.2). Many people also carried signs with these same words: but the imagery of the words 'I Can't Breathe' written across a face mask was particularly striking, signalling an uncanny consonance with our deepest fears of the pandemic: the loss of the capacity to breathe from COVID.

These political forces create the sociomaterial and economic conditions which have seen Black Americans die from COVID-19 at twice the rate of white and Asian people in the US. Indigenous Americans are also disproportionately affected, with a death rate slightly lower than Black Americans (APM Research Lab, 2020). African history scholar Daniel B. Domingues da Silva has pointed out that the expression 'I can't breathe' emerges through a long history of Black asphyxiation through extended and continuing processes of systemic and structural injustice (Domingues da Silva, 2020). Gabriel O. Apata (2020) has extended this discussion, using the concepts of air, breath and breathing to examine how racism has become 'invisible' as it has shifted from the Black body as the target of oppression to air as a vehicle for social, economic, and cultural suffocation. Through these processes, Black lives are extinguished, suggest-

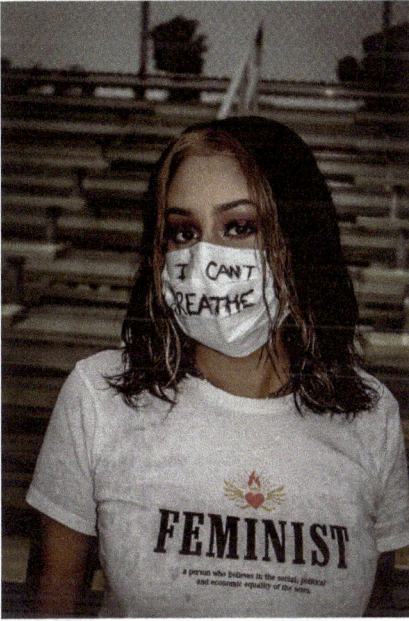

Figure 4.2: A face mask worn to protest against the death of George Floyd. Photo credit: David Ramos, Unsplash.

ing 'racial suffocation can be theorized as the slow and methodical process of social strangulation that occurs by ... the withdrawal or denial of vital socio-political and economic means of sustaining life' (2020: 246). This process works, Apata continues, by 'the application of pressure' in the form of inadequate education, lack of basic healthcare, mass incarceration, and poor housing that work together to 'make up the social air that suffocates, chokes, strangulates, asphyxiates and finally kills, not just socially but also physically' (2020: 46). These suffocating forces have come to matter specifically during the pandemic, in which Black and Indigenous people have died in devastating numbers. It becomes increasingly evident that we do not all breathe the same way; nor are we all 'allowed' to breathe in the same way or at all. Questions and concerns about breath in the COVID crisis, therefore, are not limited to processes of inhalation and exhalation, but are intricately tied to broader historical, cultural and sociopolitical forces: including entrenched racism.

Conclusion

In this chapter, we have traced the various meanings breath takes on in the context of the COVID-19 crisis. As a deeply physiological and social process, breath can in no way be understood as a homogenous or universal phenomenon. It is, in one way or another, a more-than-human life force, implicated in both living and dying. The centrality of breath in discussions of viral transmission and prevention has enabled us to 'see' how we touch each other in ways perhaps previously unimagined. Through our examination of breath and COVID masks, the human body is suddenly recognised as more porous and contaminating than we often imagine, giving itself away through its coughing, breathing and even talking. When we extend our thinking of breath to illuminate the precarity of bodily boundaries, our deep physical and social entanglements with others become more tangibly apparent. As Alaimo's concept of trans-corporeality reminds us, humans always affect and are affected by other people and by the more-than-human worlds inside and around us. The COVID crisis affects different people and groups of people differently: down to the micro-politics of how capacities for breathing freely or safely are opened or closed off. These demarcations are often made along the lines of race and class, with those who work on the frontlines, in precarious and underplayed employment, living in crowded or unsafe conditions, struggling with pre-existing chronic health conditions and with little or no access to health care among the most vulnerable. As the crisis continues, these fault lines in our society and its structures become more visible.

5 Face Mask Making Cultures

Introduction

Once it became evident that mass masking was a recommended preventive measure against the spread of the novel coronavirus, businesses and entrepreneurs spotted the market demand for inexpensive mass-produced face masks. They responded accordingly, offering for sale a diverse range of inexpensive items with a bewildering range of patterns and motifs beyond the standard blue disposable medical mask. Despite a glut of cheap options quickly saturating retail outlets, hand-crafted COVID masks also became a significant element of the broader sociomaterial landscape of masks. This chapter focuses attention on the making cultures of these types of masks. Hand-crafted masks are made by people in very small numbers from personally selected fabrics or reappropriated household items. They are sewn, glued and tied together from various and multiple materials. The skilled practices involved in their making largely fall under the broad category of craft work. Such masks, liked other hand-crafted items, are made with care to last for a long time. We position these making practices within the broader landscape of contemporary DIY cultures, focusing on the ways that the profit-resistant, creativity- and community-oriented aesthetic and political ethos of DIY was brought to bear in the pandemic, entangling with the meaning, materiality and multiplicity of the COVID mask. Drawing on social research on DIY and crafting, we argue that by thinking of mask making as a craft, we can attend to the sociomaterialities of making practices and analyse the embodied, affective and sensory work that goes into the mask itself. We discuss four different kinds of crafted masks: the artisan mask, the home-made mask, the makeshift mask and the community drive mask.

Crafting cultures

A series of COVID-related issues coalesced in 2020 which provide the context for the rise of the hand-crafted mask. Around the world, official health advice recommended, and in some places made mandatory, the wearing of face masks to slow the spread of the novel coronavirus (Chapter 1). During the early months of the pandemic, rapidly increasing demand, production delays and difficulties with cost and accessibility together limited the supply of PPE, including medical masks, within health institutions and across the public market (Chapter 2). In addition to these broader production and policy changes concerning COVID masks,

https://doi.org/10.1515/9783110723717-006

during first lockdown periods, restrictions on physical movement and shut-downs of schools, businesses, workplaces and universities limited people to their homes for extended periods of time. In response, many people in the Global North were motivated to take up various domestic crafts and creative practices (Easterbrook-Smith, 2020; Gammon and Ramshaw, 2020). The bread-making phenomenon became a symbol of this return to old-fashioned, home-based ways of spending time, with social media flooded by memes and tweets remark-ing on the apparent obsession with this activity and calling for advice on how to bake the best loaf. Supermarkets were quickly denuded of their supplies of flour and yeast as the trend caught on (Easterbrook-Smith, 2020).

It has been suggested that some people who were taking up these pursuits were seeking a form of nostalgic return to a pre-COVID past as a way of escaping the challenges of the initial months of the crisis (Gammon and Ramshaw, 2020). However, it could be argued that the turn to these activities is simply the latest development in a long-term trend towards participation in DIY and crafting cul-tures more broadly. The internet, rather than bringing analogue craft and DIY to an end, has fed the growth of practices such as home gardening, jam making, baking, sewing, art making and zine making. The contemporary turn to DIY is linked to the craft movement. Broader than the image that often first springs to mind of home improvements and renovations, DIY cultures today encompass a proliferation of diverse practices: from self-publishing to woodworking to tech-nological innovations (Ratto and Boler, 2014). The participatory and sustainabil-ity-oriented politics of contemporary DIY have become central to many creative communities. While in practice there are significant overlaps between craft and DIY practices and meanings, there are also substantial political and aesthetic dif-ferences. Whereas craft is imbued with notions of craftsmanship (Sennett, 2008), of artistry and expertise, a defining element of DIY is a celebration of the ama-teur and of participation itself (Watson and Bennett, 2020). This style and ethos have grown through the countercultures of the late twentieth century, including the punk movement, in a substantive way. Consequently, many communities of DIY practice coalesce around means and ideals of resistance, community-build-ing and radical accessibility. What cuts across the many creative media that make up DIY is a call to *join in*, using your own hands and existing abilities and the things that already surround you, with others who are also doing it themselves so that you might do-it-*yourselves* and do-it-*together* (see Ratto and Boler, 2014).

When considering the COVID mask, as previous chapters have stressed, em-bodiment and materiality are valuable concepts for making sense of the work that face masks (and mask wearers) do. These concepts are also central across sociological and geographical work on contemporary crafting practices and com-munities (Price and Hawkins, 2018). For example, Mary Carolyn Beaudry (2006), whose work spans archaeology and anthropology, charts a history of needlework and sewing through material artefacts including pins and thimbles. She interrogates existing understandings of the cultural implications and historical gendered assumptions surrounding these practices. Beyond analysis of the contexts and uses of crafted artefacts, such research illuminates the significance of understanding how and why these objects come into being.

We find Jane Bennett's work on the enchanting nature of objects (2001) and Noortje Marres' (2016) work on material participation to be inspiring approaches to understand COVID mask making cultures. Turning to questions of ethics *vis a vis* the interconnectedness of things, Bennett focuses attention on enchantment as 'a mood of lively and intense engagement with the world' to consider how this 'plays into an ethical comportment of generosity toward others' (2001: 111). The way Bennett uses enchantment to move beyond considerations of commodity fetishism is useful for considering the power of the hand-crafted face mask. Marres, a sociologist and science and technology studies scholar, explores 'public engagement as an embodied activity that takes place in particular locations and involves the use of specific objects, technologies and materials' (2016: 7). Marres' main focus, drawing from the 'post-instrumentalist' work on the politics of mattering by Latour and Bennett (among others), is on material actions of everyday practices such as making. She examines the performative, participatory modes through which public issues become located in and articulated through the material. The making and wearing of a face mask is a novel phenomenon for many people around the world, brought into the everyday only through an unprecedented global crisis. However, the labours of hand-crafted mask making are precented and mundane in comparable ways to the kinds of material actions on which Marres focuses. Marres argues 'we must study the making of participatory objects' (2016: 9).

The artisan mask

In the early months of the pandemic, when mass masking was first suggested as a way of slowing the spread of the coronavirus, traditional retailers struggled to keep up with demand. One market which thrived with options in response was Etsy (esty.com), a global online marketplace via which independent and typically individual crafters sell their handmade or vintage wares. Sellers sign up to the platform and have their own 'shop fronts' within it, paying Esty a small cut of every sale. Customers can browse through given categories or search using their own keywords. The location of every seller is prominent on their store page, and shoppers can refine the products they see by, among other sub-categories, the regions in which items are produced. Through this feature, shopping 'local' is often encouraged by design. Sellers can personalise some of the visual elements of their store page, and many share personal details about themselves and why they care about making their products. While common features of traditional retail e-commerce sites are also key for Etsy, such as quantitative rankings and reviews, these personalised, customised and 'human' elements are central to the ongoing success of the platform, just as they are to the ethos of the craft and DIY movements.

In the pre-COVID world, Etsy was a hugely popular place for finding handmade treasures from jewellery to homewares, children's toys, and wedding décor. These days, crafters have (re)turned their hands to the sewing machine. By the end of 2020, the number one selling item on Etsy was handmade face masks, with more than one million different masks available for purchase on the site. So too there was a multitude of online articles offering guidance for shoppers: from 'The 15 Bestselling Reusable Face Masks on Etsy that Customers Give 5 Stars' (Williams, 2020) to 'How to Shop for Face Masks on Etsy, According to Experts' (Horvath, 2020). As reported in this latter article, face masks generated USD133 million in sales on Etsy in April 2020 alone.

This economically successful significant phenomenon represents a desire for a particular kind of crafted face mask that has been handmade by people such as those who use Etsy. It operates at a novel intersection of digital and non-digital materialities with crafting and health risk cultures. Since the platform's launch in mid-2005, scholars interested in various aspects of digitality and craft have sought to analyse Etsy's workings. Susan Luckman's (2013, 2015) scholarship on Etsy explores each of these elements and raises important considerations relevant to the case of the face mask. She focuses on the relationships between on- and offline crafts markets to critically consider the 're-articulation' and 'return of credibility' to domestic craft practices: in particular, those involving yarn and sewing. Tracing a history of British women's craftwork during the time of the Industrial Revolution to consider the societal shifts comparable with our contemporary moment, Luckman argues: 'Today, when direct connections to the hands that produced the goods we own are rare, an abundance of mass-produced goods reinstates a Benjaminian aura to the analogue and the handmade' (Luckman, 2013: 264). What Luckman speaks of as an aura is similar to Bennett's understanding of enchantment. Enchantment for Bennett (2001: 5) is 'a state of wonder ... a momentarily immobilising encounter' where you are

> brought to rest, even as the senses continue to operate, indeed, in high gear. You notice new colors, discern details previously ignored, hear extraordinary sounds, as familiar landscapes of sense sharpen and intensify. The world comes alive as a collection of singulari-

ties. Enchantment includes, then, a condition of exhilaration or acute sensory activity. To be simultaneously transfixed in wonder and transported by sense, to be both caught up and carried away – enchantment is marked by this odd combination of somatic effects. (2002: 5)

Bennett's description of the enchantment of things outlines a more eclipsing experience than one typically has while clicking through the store pages of Etsy's online marketplace. It is nevertheless a useful concept with which we can extend the work of Luckman's aura and make some sense of what matters in the Etsy COVID mask phenomenon. Both Luckman and Bennett consider ethical orientations and engagements of human-nonhuman assemblages by identifying the ways certain artefacts unearth and engage the structural injustices of environmental, economic and humanitarian crises. Bennett argues these points most explicitly, stating that 'enchantment is a mood with ethical potential' with which we can centre 'the question of motivation, the "how" of ethics' (2001: 131). The question of ethics is a key element in the contemporary desire for local and handmade goods. This shift is one of consumer power, away from the environmentally and economically damaging mass production of goods. As Luckman (2015) contends, such discourses of ethical consumption fuel Etsy's significance.

For Etsy COVID masks, aesthetics is as central in their enchanting aura as the mode by which they are produced. As Figure 5.1 shows, on Etsy and other handicraft domains, the range of COVID masks available is extraordinary. No matter the print, most masks are labelled 'natural', 'washable', 'reusable', 'organic' or 'stylish'. As is the norm with products sold on Etsy, it is presumed that these masks are made by individual crafters who are working solely on small and limited runs of goods. Their design, and the fact they are sold via Etsy, mean these masks read as if they are exclusive *and* hand-crafted. Multiple elements make these masks local, artisanal, ethical and therefore desirable as *not* mass produced. Hand-dyed cotton masks in earthy tones, mask and hair scrunchie sets sewn from matching fabric, brightly coloured tie-dyed masks and various floral prints are popular. You can imagine the fabric for each mask being selected, the crafter's fingers running over rows of material until they land on the right one: the same fingers that later sew the masks carefully, sketching the right lines in pencil and pushing pins into tracing paper, scissors separating and thread holding everything together.

There is a mask for every occasion and mood, to match any outfit. Customised masks for brides and bridal parties can be found, beautifully embellished with lace, seed pearls or diamantes (see Figure 5.2 for an example of such a mask). Masks are available for sports team supporters and for every milestone in a person's life: christenings, bat and bar mitzvahs, birthdays and funerals,

Figure 5.1: Hand-crafted COVID masks. Photo credit: Gabriella Clare Marino, Unsplash.

Figure 5.2: Artisan mask as part of an Indonesian bride's outfit. Photo credit: Ahmat Muhlisin, Unsplash.

with appropriate messages and images to mark the occasion. Various national themes feature in masks: not only national flags (Chapter 2), but for example, Australiana such as colourful illustrated fabrics with wattle flowers, gum leaves, koalas and kangaroos with joeys in their pouch. Perhaps the most self-consciously 'arty' of all COVID masks are those offered by several art galleries and museums across the world, including the London-based National Gallery and Tate Museum, the Rijksmuseum in Amsterdam, the Klimt Villa in Vienna and the Metropolitan Museum of Art in New York, which have begun to offer masks depicting some of their most well-known masterpieces in their online shops. Taking this idea even further, the Fitzwilliam Museum in Cambridge, UK, offers greeting cards for sale in which some of its famous portraits by artists such as Millais and Titian have been 'reimagined' with the addition of face masks on the subjects (Woodyatt, 2020).

Bringing the artisan crafted COVID mask into being, therefore, is a confluence of existing and emergent structures which are cultural, material, economic and affective. These include e-commerce, a pandemic and ensuing public health directives, retail shortages, the visibilising work of social and online media, and the contemporary turn away from the mass-produced anomic goods of late capitalism toward a desire for the boutique, the personal, the authentic and the artisanal. With and through these elements and agents, the style of the mask and style of the making are intra-actively co-constituted to generate enchantment, tempting buyers to purchase.

The home-made mask

Enchantment does not only pertain to a mask buyer's engagement with the mask artefact, however. Luckman draws on Bennett's work on enchantment to argue that in reference to craft as a *process* specifically, making can also be an experience of 'enchanted engagement' (Luckman, 2015: 79). A number of domestic items became effectively impossible to find and purchase at different times in 2020: not only flour and yeast, but also jigsaw puzzles, seedlings, bicycles and guns. Some items, such as toilet paper and hand sanitiser, sold out because of mass 'panic buying' when the coronavirus was officially declared to be a pandemic. Other items, such a puzzles and pasta ingredients, became scarce as people sought activities that would fill their long housebound days under lockdown or during extended periods of unemployment. The swift and widespread uptake of domestic craft and DIY projects, and the difficulties of finding suitable or desirable face masks as traditional suppliers struggled to meet demand, resulted in a shortage of fabric and sewing machines in many countries (Kavilanz, 2020). In addition to the surge of crafted face masks sold on platforms such as Etsy, many people purchased sewing machines to make reusable face masks for themselves and their loved ones. This gave rise to another novel face mask phenomenon: the home-made mask.

As the spread of the coronavirus became more serious. so too did the need for effective face coverings. Sewing patterns for face masks quickly spread across the internet. Many retailers, such as Australian craft giant Spotlight, shared official patterns for free online. Others, made by independent crafters, shared their own designs via social media. However, many people who were interested in making their own masks did not have the necessary knowledge and skills to simply put a pattern to work. A new form of the highly popular online tutorial genre quickly filled this gap. A cursory Google search for 'how to sew a face mask' returned over 1.1 million results at the time of writing; 'how to make a face mask' resulted in more than 5.4 million results, with almost 1.2 million of those in video format. In our contemporary era, the YouTube 'how to' video is a well-established and highly lucrative creative medium (Burgess and Green, 2018). The youth-oriented content sharing platform TikTok, which received a massive increase in users during 2020, also presented many short-form videos about home-made face masks and cultivating sewing skills for beginners.

As with other online tutorial videos, in many guides for making DIY face masks there is a tension between the amateur and the professional. This tension is particularly important in the case of COVID masks, where technical considerations about the best fabrics to use and how many layers to include in the mask are crucial to whether the mask will be effective in achieving high levels of protection against the spread of the novel coronavirus rather than simply look decorative. Etsy directly acknowledged this issue. In mid-April 2020, the platform put together a guide for people selling masks via their site which included the advice that: 'You shouldn't make any medical or health claims about your products, even if you believe they're true. Because of this, you should avoid mentioning COVID-19 or coronavirus anywhere on your mask listing pages' (Etsy, 2020). It is important to note here that such advice does not mean that crafters and DIY makers are not carefully seeking out guidance and information that will make their masks more suitable with regards to preventing the spread of the coronavirus. In fact, the intensity of care and consideration that has gone into the making of masks is one reason why the hand-crafted version has such an aura.

By mid-2020, thousands of feel-good stories circulated on official and social media channels of people making face masks at home which they then gifted to their families and friends, some of whom were separated by oceans and closed international borders, and others who they could not see in-person despite living only streets away (Haggerty, 2020). Thought and love were deeply given to the selection of soft, breathable and washable materials, to the packaging and other care items that masks were sent with, to the surprise which accompanied their posting. Bringing such elements into account, the sharp rise in DIY mask making speaks to more than the practical necessity for people to locate face coverings at a time when they were hard to purchase through typical avenues. The American Association of Retired Persons (AARP), for example, an American not-for-profit platform dedicated to the health and wellbeing of people over the age of 50, published an article on making masks. They reported on an interview with a 'crafter', a woman who has been 'sewing face masks as quickly as she can turn them out' while furloughed from her workplace; the reporter writes, this gave 'her time to sew – something that feels tangible in a time of uncertainty' (Leach, 2020).

A fundamental quality of digital platforms such as YouTube and of DIY communities such as zine making are what scholars call 'participatory culture' (Burgess and Green, 2018; see also Watson and Bennett, 2020). Participation is key to the home-made face mask. The style of the 'how to' guide, in terms of the language of delivery and the (lack of) existing knowledge such guides assume, cultivates an invitation for participation which is intentionally and obviously open to any and all who are interested. There are no (or very few) boundaries limiting access. Anyone, these guides imply, can make a mask and put their own hands to work to create something suitable-enough in this time of global crisis. Further, as Marres' (2016) work on participation emphasises, it is also important to consider this mass crafting initiative as an event of individuals discretely working in their separate domestic DIY bubbles. It is important to question 'how things mediate publics' because such questions shape 'how we understand the affordances of everyday settings for engagement' (Marres, 2016: 24). The DIY making of COVID masks is a crucial site for public engagement in the social crisis of the pandemic.

The makeshift mask

As the rush on sewing machines partly shows, many keen DIY mask makers did not have adequate materials on hand. In response, a plethora of online guides have been made specifically for no-sew face masks. Old socks, T-shirts, paper towels, bandanas and pillowcases have all been identified as feasible starting materials which require little beyond safety pins, scissors and elastic bands to become COVID masks. Across different online media, from blogs to TikToks and zines, accessible and easily replicable guides rapidly spread as the seriousness of the pandemic became globally apparent in early to mid-2020. As well as sharing people's lived experiences of the pandemic (Gharib and Harlan, 2020), many zine publications in particular have also focused on mutual aid; they share information on the COVID crisis itself as well as tips for social distancing, making masks at home, and how to set up DIY emergency hand washing stations in public spaces for people in need such as unsheltered refugees (Cheung, 2020).

Interestingly, many online video and blog guides for makeshift masks include numerous caveats and legitimising references. The blog *Sarah Maker* (2020), for instance, links to a research article and notes that the study 'found that cotton T-shirts and cotton pillowcases are the best at-home materials for making DIY face masks, due to their ability to capture small particles yet remain breathable'. The same article also notes that 'A DIY face mask is not a replacement for a surgical mask' and 'masks are not a replacement for social distancing'. Such lines, commonly used to frame advice for how to make DIY face masks, serve as protective provisos for guide publishers and are also examples of how people with online platforms engage networks of expertise to share information and skills with others.

Adding another layer to the landscape of circulating hand-crafted COVID mask information was the official advice given out by numerous government health organisations. This advice boils down to the idea that any mask is better than no mask. However, there are differences between the advice that countries officially give. The CDC has pages on its website for selecting, wearing, making and washing face masks. On the making masks page, the CDC includes text, image and video-based guides for both sewn and no-sew face masks from two layers of cotton fabric (Centers for Disease Control and Prevention, 2020d). Public Health England offers very similar advice via images and text, for making a two- or three-layered mask from cotton using a sewing machine (Public Health England, 2020). The Australian Government Department of Health instructions are for making masks with three layers (Australian Government Department of Health, 2020a), and different kinds of material are recommended for each layer – water-resistant, blends, and water-absorbing. Notably, the Australian advice names fabric types under these material categories as well as common household materials that could also be used, such as exercise clothing, reusable 'green' shopping bags and shoelaces for ear loops. Such advice, it is critical to highlight, is given by such institutions within the context of the severe undersupply of medical masks for people working in health services or other front-line roles (Chapter 2). The CDC, for example, stresses in bold at the top of its information pages: 'Do NOT use a mask meant for a healthcare worker. Currently, surgical masks and N95 respirators are critical supplies that should be reserved for healthcare workers and other first responders.'

Makeshift masks have a distinct affective presence. Unlike the curated e-commerce stores of the artisan masks, or even the crafty and almost-kitsch tutorials shared for home sewers, information about these kinds of makeshift masks predominantly come from official government webpages which make strategic use of infographics and public health advice. Such texts significantly differ in tone and intent. As noted through the Australian example, this is partly because makeshift mask guides appeared later than the initial Esty COVID mask boom, for example, as the official advice on wearing masks changed. This advice also stems from the urgency that people were feeling in finding it difficult to source face masks which they could order online to be delivered to their homes. As a result, while they are similar in practical intent to the artisan or handsewn masks discussed above, no-sew masks are arguably *dis*enchanting. Cobbled together from socks, elastics and old T-shirts, these masks generate a disenchanting mundanity which is striking in a context of increasing seriousness and worsening crisis.

In this way, makeshift masks can be seen as almost a generative inversion of the conditions of enchantment which Bennett (2001) takes as her focus. Nonetheless, such masks may offer those who make them a feeling of purpose, pleasure and achievement in the very act of transforming one mundane object into a useful (even lifesaving) thing with little more than sleight of hand, thereby recycling things that otherwise may be destined for landfill. Indeed, Bennett's later work on thing-power in her book *Vibrant Matter* (2009) draws attention to the force and vitality of assemblages – their thing-power – that may be perceived as waste or garbage. Affective forces can be intense in the reappropriation of strikingly mundane materials in a world of crisis and dearth of needed materials. This affective intensity opens an engagement with the distributions of responsibility across the provision of masks for and of health and care and returns our attention to some of the ethical complexities of care surfaced by more-than-human scholars such as Bennett (see Chapter 6 for further discussion of these issues).

The community drive mask

When the WHO released a statement warning of the impacts of the short global supply and significantly increased costs of PPE for healthcare workers (World Health Organization, 2020b), it called for industry and governments to significantly increase manufacturing and manage supply chains. Communities around the world also responded. Mass numbers of community drives were organised to meet the demand for COVID masks across the healthcare sector and across society more broadly. Thousands of people joined sewing bees, making masks together from their own homes. These collective efforts were different from the masks people sewed for themselves and their families and friends or sold on sites like Etsy. They involved the coming together of skilled and novice strangers to make masks for free for other strangers, typically for local or national groups of people in need who were particularly vulnerable due to the wider circumstances of the pandemic.

The Morwell Neighbourhood House in the Australian state of Victoria is one example of such an initiative. This centre offers programs, activities and support services for people in the area facing poverty, distress, disadvantage or isolation. Masks were recommended and then made mandatory by the Victorian government from July 2020 as a second wave of COVID cases rose across the state. Morwell Neighbourhood House used their social media channels to call for volunteers to make and donate face masks, reportedly hoping to recruit ten volunteers to make 100 masks in total which would be freely distributed to people in need in the local community. Many more people than expected responded to the call, with 80 volunteers joining the sewing bee within two weeks (Whittaker, 2020). The centre was then able to supply hand sewn COVID masks to hundreds of people who were on a pension, living with a disability, immuno-compromised or otherwise disadvantaged: masks made with care by another member of their community.

Larger campaigns have also arisen in countries particularly hard hit by the COVID crisis, including the USA and the UK. In the USA, one of the most significant community sewing drives has been the 100 Million Mask Challenge (American Hospital Association, 2020). Initially launched across the west coast of the country by Catholic health care provider Providence, before being expanded to a national scale by the American Hospital Association, this community drive notably brings together individual mask makers with a network of community organisations and local manufacturers to address the severe undersupply of PPE across the country's health care system. Additional resources are provided on their website, including sewing patterns and video tutorials for making masks. Part of the effort includes a 'Wear a Mask' campaign (a direct rejoinder to anti-mask movements in the USA) and a campaign for people to donate money to their local hospitals. Also connected to this initiative are community-based organisations such as emergency housing shelters and food banks, to which people are encouraged to donate masks and other needed short-supply items such as disinfectant. The Providence website additionally expands its call for care, listing other ways that people can help at this time (Providence, 2020). This includes donating blood, delivering groceries and other goods for more vulnerable neighbours, and telephoning friends to check in on them and curb social isolation.

In the UK, the Big Community Sew is a similar drive for home-sewn face masks that brings together individuals and local community sewing groups. Launched by Patrick Grant, a famed Scottish fashion designer, the UK-wide effort aims to ensure 'every person in every community in Britain has the face covering they need' and asks people to 'Find out who in your community needs face coverings, and get making' (The Big Community Sew, 2020). This drive is grounded in a simple yet compelling message: 'There are around six million sewing machines in homes across the UK and if every one of those machines can be used to make just a dozen face coverings that would be enough for every person in the UK'. Like the 'Wear a Mask' campaign linked to the US-based 100 Million Mask Challenge, the Big Community Sew also incorporates a competition called #gotitcovered, aimed at encouraging people to participate in the campaign, to wear masks in public, and to not use medical masks which are needed by healthcare workers. The Big Community Sew website hosts instructions for making two kinds of face masks, with patterns to fit adult and children's faces, and includes information on making masks from common household items such as pillowcases.

The Big Community Sew website also hosts a series of tutorial videos featuring well-known and expert fashion designers sewing brightly coloured COVID masks in their homes. People are encouraged to make 'creative', 'sustainable' and 'well-fitting' masks and share their creations on social media using the competition hashtag. New and existing community sewing groups are also able to publicly sign up to the drive and be listed on the website, to encourage other interested local members to join their efforts and to help with fundraising. Like the individual bloggers and artisan mask makers discussed earlier in this chapter, the Big Community Sew website also draws on official advice from several UK-based public health agencies and shares related up-to-date information. This includes very clear advice on how to wear a face mask properly – 'The important thing is that it should cover your mouth and nose while allowing you to breathe comfortably' – and why people should wear a mask:

> They are not intended for the personal protection of the wearer – they are designed to prevent people who have COVID-19, but might not know it, from spreading it to others. In simple terms, if I wear one I protect you, if you wear one you protect me. (The Big Community Sew, 2020)

It is notable that efforts like the Big Community Sew stress community care as well as sustainability. Such efforts can be seen as 'public demonstrations of the "thing-powers" of engagement' with which we can be engrossed and implicated (Marres, 2016: 97). These intensities, vitalities and relational connections are significant for each of the DIY masks discussed in this chapter: that people, living through a pandemic, go to the trouble to make something themselves that could help others or opt for wearing reusable home-made masks rather than the disposable or mass-produced masks made by large companies finding new ways to make money. The hand-crafted face mask and the practices and materialities of public participation amidst the COVID crisis expand our visions of care beyond the self to the extended social and more-than-human world.

Conclusion

As Marres discusses in her work on environmental technologies, objects and devices enable (and 'are presented as enabling') a 'material form of participation' (2016: 67). This is clear in the case of the crafted COVID mask. Through the careful and care-full arrangement of materials in and of the personal space, the practices of COVID mask making configure public participation as a form of sociomaterial action. This action involves engaging in ethical care for others by making masks for them. The making, sharing and wearing of COVID masks is recognised by people as something productive and meaningful to do amid uncertainty and risk. An ethos of anti-consumerism, community support and care is made material through the hand-made mask. The making of masks, as well as the wearing of masks, are not only symbolic of these values but are practices through which people may materially participate in a caring COVID society.

6 Face Masking and Care

Introduction

Care has become a major issue in the wake of the COVID crisis in a multitude of ways. The COVID mask has become a focal point of discussions of care: portrayed not only as a protective apparatus for those wearing it, but even more as a device to protect others. As we noted in Chapter 3, people are encouraged to follow specific guidelines for caring for their reusable masks so that they are safe and effective as protective agents. And in Chapter 5, we drew attention to hand-crafting COVID masks as a practice of care for others. There has been a heightened focus on the impact on healthcare systems and healthcare workers by people seeking testing or medical care, as well as an emphasis on how individuals should engage in self-care and informal care for others outside the medical setting. At the same time, those who refuse masks are positioned as lacking care or even potentially hostile to the health and wellbeing of others. In this chapter, we further consider these dimensions of face masking and care. We engage with theories of care to reflect on how masks and mask wearing and creation become implicated in the ethical, political and material dimensions of care for both human and nonhuman others. Locating ourselves in the rapidly shifting and emerging conditions of the COVID pandemic, masks become central to our ethical and careful responses.

Philosophies of care

In the age of COVID, the simultaneous individual/community concept of health and illness returns us to earlier ideas of the body, existing before the advent of individualised healthism and the focus on lifestyle in the mid-twentieth century. Pre-modern concepts of embodiment positioned individual human bodies as more open to the world than contemporary western understandings, with continuous outbreaks of infectious diseases a constant warning that people were highly vulnerable to unexpected contagion (Shildrick, 1997; Burfoot, 2010; Cole, 2010; Lupton, 1995, 2012, 2021). By the late twentieth century, the WHO and other authoritative health bodies were proclaiming that due to improvements in sanitation, universal healthcare and the development of antibiotics and vaccines, infectious diseases had been well contained and controlled in many regions globally. Outbreaks of novel infectious diseases, beginning with Ebola virus disease in the 1970s and continuing with HIV/AIDS in the early 1980s, fol-

https://doi.org/10.1515/9783110723717-007

lowed by SARS, MERS, Zika virus disease, swine influenza and avian influenza, have emphasised the role of nonhuman agents such as viruses and bacteria in human health and wellbeing globally (Lupton, 2021).

These changes in conceptualising and responding to health threats are accompanied by transformations in notions of care. There are various philosophical ways to understand the concept of care. Foucauldian theory has emphasised the importance placed on people in late modern wealthy countries to care for the self. In his book *The Care of the Self* (Volume 3 of his *History of Sexuality*), Foucault (1986) emphasised that the concept of the care of the self stems from ancient Greek and Roman philosophies. They describe the care of the self as an ethical practice that involves cultivating knowledge of the self and knowledge of rules of conduct. The care of the self is both a duty and a set of practices (technologies of the self) in which people engage to achieve health and wellbeing. Foucault argued that this ethic of practising the care of the self remains a dominant mode of selfhood and embodiment in contemporary societies. Crucially, he emphasised the material dimensions – the technologies or practices – of the care of the self. Foucault argued that rather than people being coerced into adopting these practices, they did so willingly, in their own interests.

Critical care theorists take a more macro-political approach to understanding the power relations that are part of care arrangements and practices, as well as the human rights issues of those who need care (Tronto, 1993; Fine and Tronto, 2020). This approach to care highlights that care often involves unappreciated acts of maintenance and repair to support human wellbeing and flourishing (Tronto, 1993). Critical care theorists point out that care for others has traditionally been feminised and devalued: positioned overwhelmingly as the responsibility of women and people of colour, particularly in informal care settings such as the home. A critical care approach is sensitised to the relations of power involved, and the ways in which caring can be paternalistic, colonising, sexist, racist, exploitative or otherwise restrictive of people's agencies. From this perspective, it is argued that those in positions of power and privilege tend to benefit from the care labour offered by others, with little expectation that they reciprocate care. Care work remains largely invisible and underpaid (or indeed, often unpaid), viewed as requiring little skill or training (Tronto, 1993; Fine and Tronto, 2020). In emphasising the webs of connections that link people to each other through acts of care (Tronto, 1993), this approach has a relational yet distinctly humanist perspective.

Other critical approaches include both a focus on power relations and the role played by material objects in care. The sociology of health literature has included many sociomaterial analyses of medical technologies and other objects contributing to healthcare (such as bandages, pharmaceuticals or medical records) (Martin et al., 2015; Buse et al., 2018; McDougall et al., 2018). Drawing particularly on Latourian actor-network theory, these analyses have highlighted what has been referred to in some of the literature as 'materialities of care' (Buse et al., 2018) or 'care assemblages' (McDougall et al., 2018). This approach to care assemblages have been taken up by some feminist science and technology studies scholars investigating the ethics and politics of care. They ask who benefits from the use of material objects in care and how these relations might be otherwise: who is the object of care, how to care, what are the politics of care, who has the power to care? (Martin et al., 2015).

More-than-human perspectives on care drawing on Indigenous/First Nations and feminist new materialism theory also identify that caring is not simply an act of people with other people, but always involves other entities. From this perspective, it is important to identify relations of care (Puig de la Bellacasa, 2012, 2017; Braidotti, 2020; Hernández et al., 2020). This approach recognises that agencies are generated and shared with and between the agents in care assemblages. It acknowledges the more-than-human dimensions of care as it emphasises the co-constitutive entanglements between humans, nonhuman others, objects and environments. Critical and more-than-human perspectives seek to identify the ambivalences and difficulties of care. Being the recipient or bestower of care is not always a harmonious, benign experience. It can create disturbing and disruptive affective forces. 'Thinking with care' (Puig de la Bellacasa, 2012) means being alive to these affective and relational tensions and conflicts at both the micro-political and macro-political levels.

The COVID crisis has emphasised the structural inequalities that persist globally, demonstrating more than ever that the archetypal privileged human subject is white, male, young, able-bodied and heterosexual. It is this subject who is largely cared 'for' and 'about' in COVID-related care. This subject is not generally forced to continue to engage as a carer or in other 'essential services' that are low-skilled, under-paid and which expose workers to greater risk of contracting the novel coronavirus (Bambra et al., 2020). And it is this subject who feels free to reject face mask wearing as an act of weakness and a mark of stigma and shame, demonstrating little concern for others he may expose to risk (Capraro and Barcelo, 2020). President Trump is the most well-known exemplar of this privileged subject.

Face mask wearing as symbolic of care for others

Surgical-grade face masks as well as the other PPE worn by healthcare workers (such as face shields, gloves, gowns and full hazmat suits) quickly become potent symbols of the 'frontline' of the fight against COVID. Some news stories featured photographs of the weary faces of healthcare workers, masks removed, with the deep red marks inscribed on their flesh from hours of mask wearing demonstrating the embodied discomfort they bore in working on the frontline for hour after hour. Thus, for example, an article published in *The Atlantic* in March 2020 featured images taken of nurse and doctors working in a hospital in the small Italian city of Pesaro (Weiss-Meyer, 2020). The article was headlined 'The visible exhaustion of doctors and nurses fighting the coronavirus', with the subheading, 'Documenting the marks the pandemic is leaving on medical professionals in Italy'. Italy was one of the hardest hit countries in the early stages of the COVID crisis. Each of the healthcare professionals featured in the story is shown in their PPE apparel but most have removed their face masks to show their whole visage. These photographs are used to convey the massive strain the Italian healthcare system, in general, was under at this point in the pandemic: not just in this hospital in this small city, but across the entire country: 'When they finish their shift and remove their mask, they bear deep imprints – physical and emotional – of their efforts to ease the crisis' (Weiss-Meyer, 2020). The dramatic images of frontline workers with sore and damaged skin attracted the attention of the public, health organisations and governing bodies, as well as commercial companies.

In addition to painful bruising and ulceration from extended mask wearing, some healthcare workers also experienced 'maskne', the term given to the phenomena of skin irritation and acne flare ups as a result of mask wearing (discussed in Chapter 3). In the USA, the American Academy of Dermatology Association (2020) responded to these emergent issues by creating an official online resource document titled 'Skin protection guidelines for using PPE' to help health professionals treat and prevent skin conditions caused by masks and other PPE. Personal care and beauty brand Dove (2020) also responded by launching a global campaign called #CareFromDove. The campaign donates personal care products and protective equipment to government and NGO initiatives around the world, including products that can help protect skin from irritation and damage resulting from intensive mask-wearing. At the time of writing, Dove claims to have donated over $5 million US globally and maintains their efforts will continue. The #CareFromDove campaign, therefore, emerges as a material and public expression of care for frontline workers, albeit one entangled with capitalist and economic interests.

Informal crowdsourced resources also emerged as images and personal stories accompanied by the hashtags #soreskin and #PPE circulated on social media platforms. In October 2020, a registered nurse from Boston issued a tweet expressing feelings of deep exhaustion and physical pain from the ulcers developing on her nose from long hours in a surgical mask. The tweet was retweeted almost 80,000 times, prompting a flurry of responses from others around the world. Many of these responses offered expressions of support and solidarity from other frontline workers, often through the sharing of personal tips and strategies for dealing with similar issues such as the use of specific products or 'hacks' that proved to be particularly helpful for protecting the skin. These virtual offerings emerge as performative materialities of care and social media platforms and expert and informal knowledges emerge as integral to the PPE-care assemblage. Mask wearing as a sociomaterial practice of care proliferates in unexpected ways, producing new relationships between humans, objects, and technologies that are embedded within specific capitalist and consumer networks.

The surgical face mask as part of the PPE assemblage was also used in promotional materials calling for the support and protection of healthcare workers. For World Health Day 2020, the World Health Organization distributed a poster showing a healthcare professional swathed in PPE (including a surgical mask and face shield) and surrounded by two large, stylised images of the novel coronavirus. The text read: 'Nurses, midwives and other health workers are at greater risk of COVID-19 than anyone else … thank them for their bravery and hard work'. Another awareness campaign run by the Laura Hyde Foundation, a charity to support the mental health of healthcare workers, used images of people in surgical attire and wearing masks with the words 'There's no mask for mental health'. The message of this campaign was that masks could only protect against certain health problems, while seeking to emphasise the toll that the COVID crisis was having on healthcare workers.

Other public health campaigns focused on the importance of respecting and caring for one's family or community. A campaign based in Santa Barbara County, USA, entitled 'Protect. Respect. Wear Your Mask.', used posters and a television advertisement showing photographs of diverse individuals from this community, each identified with their name and role in the community, wearing the same type of disposable mask. Quotations from each individual (some in Spanish) included: 'I wear a mask for my family', 'I mask up for a safe + well community', 'My mask protects you, and yours protects me. We're in this together,' and 'My family wears and sews masks to keep our neighbors and friends safe'. A similar campaign in Utah used the words 'I do it for …' handwritten across disposable masks, with each individual photographed wearing a mask having a different message: for 'my friends', 'my Mom" and 'my kids'. Cornell University's campaign had photographs of young people wearing branded Cornell T-shirts and in plain white fabric masks, with the messages 'Respect your friends. Wear a mask.' and 'It's up to all of us to prevent the spread of COVID-

Figure 6.1: The face mask as symbolic of caring for friends. Photo by pixpoetry, Unsplash.

19'. A Wisconsin Council of Churches poster used stylised images of people wearing masks and drew on Christian tenets: 'Love your neighbor. Wear a face mask.' Such sentiments of care are illustrated in Figure 6.1.

The idea of embodied sociality is embraced in these portrayals of care. Human bodies are understood as interembodied and communal as well as individual (Bracken, 2020). While individual practices of self-care are still important in avoiding infection, infectious diseases are not 'lifestyle' diseases that are contained to the individual body. For outbreaks such as the coronavirus pandemic, I pose a risk to you: you pose a risk to me: together, we must engage in behaviours that reduce this risk among the body of the community. As many COVID public health messages put it: 'We are all in this together'. Both the risk and the responsibility to contain the risk are shared across the communal body. This means that efforts to contain the spread must isolate individual bodies from others: as in the measures of quarantine, self-isolation and shielding at home. People must follow individual practices of hygiene: containing their sneezes and coughs in tissues or elbows, staying at home if feeling ill, washing their hands constantly, avoiding touching their faces. But these practices are modes of containment and prevention that reduce the overall risk to the population as well (Bracken, 2020).

Politicising mask wearing as an act of care

Wearing a face mask in public (outside the healthcare context) has in some cultural contexts become a marker of difference and opened people to attack. Since the advent of the COVID crisis, Asian people who look Chinese living in the USA have reported many incidences of being made to feel uncomfortable by being shunned or avoided or even racially attacked by other Americans: especially when they were wearing masks (Zhou et al., 2020). These experiences have been reported to and documented by the Stop AAPI (Asian American Pacific Islander) Hate website (Stop AAPI, 2020), established specifically for this purpose once it was realised how COVID had intensified racism against Asian Americans in the USA. This information details how Asian Americans have been simultaneously blamed for the global spread of COVID but also mocked for engaging in a practice designed to reduce the spread (mask wearing). President Trump's continual references to the 'Chinese virus' and the 'Kung Flu' in his public appearance has contributed to these portrayals (Stop AAPI Hate Campaign, 2020). Anti-Asian racism incidents were also reported in a Pew Research Center survey conducted in June 2020 (Ruiz et al., 2020). Asian adults included in the survey noted that racist slurs or jokes directed at them and people acting as if they were uncomfortable in their presence had increased since the beginning of the COVID outbreak. They were far more likely than white respondents to report feeling afraid that they might be physically attacked or threatened by others. Both Black (42%) and Asian (36%) respondents were far more worried than white respondents (5%) that other people might be suspicious of them because of their race or ethnicity if they wore a face mask in public.

It is not surprising that Asian Americans were concerned about the threat of violence when wearing a mask. News reports in US outlets described incidents of Chinese and other people of Asian appearance wearing masks being beaten in violent attacks. For example, a *Newsweek* story reported an attack in which an Asian woman wearing a mask in a New York City subway station was set upon by a man, who yelled that she was 'a diseased bitch' and hit her with his fists and the umbrella he was carrying (Palmer, 2020). And it is not only people of Asian appearance who have been subjected to mask-related violence in the USA. In May 2020, a security guard at a store in Flint, Michigan was fatally shot after he denied entry to a customer not wearing a mask, while in October 2020, an 80-year-old man died in a bar in West Seneca, New York State, after having been shoved to the ground and hitting his head on the floor during a confrontation with another man who was not wearing a mask. The attacker became enraged after the other man had confronted him about not wearing a mask (Staff and agencies in West Seneca, 2020).

Equally powerful affective forces are evident in popular cultural portrayals of people who refuse to wear a COVID mask. Social media platforms are saturated with acerbic and comically intended condemnations of such individuals and their performances and expressions of this refusal. Many of the circulating memes use the example of the cultural figure 'Karen' who has come to represent the entitled white (often passive-aggressively racist) individualist woman who becomes exceedingly and demonstrably upset when things do not go her way, possessing little capacity for self-awareness. In the example of the 'Bunnings Karen' (discussed in Chapter 2), resistance to mask wearing was clearly performed but was met with serious backlash by social and news media. As was evident in the dominant response to 'Bunnings Karen', memes and other social media imagery tend to portray and shame individuals who do not wear masks as selfish, careless and as not caring for others. In contrast, those who do wear masks are applauded as virtuous, self-sacrificing and recognised for their ability to care for (seemingly all) others. The act of wearing a mask seems to overlook and override any potential shortcomings of the wearer, and issues like racism are erased or eclipsed. On the other hand, shame is deployed to paint and position those who do not wear masks in broad brush strokes as selfish, unintelligent, and sometimes racist. This polarising framing of care and carelessness then circulates in broader discussions of mask wearing. An article appearing on Politico.com sums up this divisive framing: 'Wearing a mask is for smug liberals. Refusing to is for reckless Republicans' (Lizza and Lippman, 2020).

Implications for the care of nonhuman others

The COVID crisis has prompted a dramatic increase in the production of single-use medical masks, with one estimation projecting monthly usage and consumption of upwards of 129 billion masks globally (Prata et al., 2020). As a response to the crisis, these masks exacerbate the very conditions that led to the outbreak of COVID. While COVID masks can play an important role in protecting humans against the spread of the coronavirus, their production, consumption and disposal hold implications for the health of nonhuman others. This includes the environment and ecosystems as well as other animals and their habitats. Disposable masks are manufactured from non-woven materials, comprised of plastics such as polypropylene, that can persist in the environment for decades or even centuries as harmful micro-plastics (Fadare and Okoffo, 2020). Although effort has been made on the part of government and health authorities to educate people how to wear masks properly, less attention has been paid to ensuring their proper disposal. Consequently, masks have emerged as a new source of pollution,

Figure 6.2: A discarded disposable face mask becomes litter. Photo credit: Cate Bligh, Unsplash.

similar to disposable nappies or wipes, with the sight of carelessly discarded masks on footpaths and green spaces becoming increasingly commonplace (see Figure 6.2).

The World Health Economic Forum (2020) has estimated that approximately three-quarters of used disposable COVID masks will end up in landfills or in the ocean. There have also been concerning reports of face masks appearing as flotsam and jetsam in oceans, lakes and rivers (Kassam, 2020). As early as February 2020, conservation and advocacy group OceansAsia (oceanasia.org) posted a photo of dozens of surgical masks discovered on Hong Kong beaches while conducting a year-long research project into marine debris and micro plastics. After finding alarming numbers of masks on multiple local beaches, co-founder Gary Stokes voiced concern that masks were adding to the already dangerous volume of plastics in the ocean, increasing risk to marine life (Stokes, 2020). Improperly discarded masks can contribute to malnourishment and illness in terrestrial and marine creatures that are unable to distinguish plastic from prey or other proper food sources (Roberts et al., 2020). Animals can also become tangled in the elastic ear loops and ties either on land or in the water, and masks that end up in the ocean (either through careless litter or landfill) contribute to the waste crisis currently choking our oceans and other bodies of water (Aragaw, 2020).

These effects of the increased use of disposable masks during the COVID crisis have garnered increased public and scholarly attention (Adyel, 2020). An article appearing in *Slate* magazine written by archeologist Pamela Geller (2020) predicts that the dramatic increase in the production and consumption of single-use medical masks brought on by the pandemic will leave an indelible mark on the future of our planet. She speculates, 'Centuries from now, archaeologists may unearth a stratigraphic layer of plastic waste – buried in landfills, scattered across landscapes, preserved within the digestive systems of extinct species'. Geller expresses further concern over the high levels of micro-plastics that will result from the slow breakdown of surgical masks on land and in water. She explains that while micro-plastics disappear from sight, they never truly leave the environment. Instead, they insidiously and invisibly make their way into water and food sources, with devasting consequences for human and nonhuman health.

While the pandemic continues, face masks and other disposable PPE such as gloves remain an imperative part of human-oriented care and protection and a focus of public health discourse. Consequently, new strategies for managing the environmental impact of single-use masks and other forms of PPE are needed that consider both human and nonhuman worlds and our entangled, uncertain futures. We are moved to question how we might produce, use, and dispose of face masks (and other PPE) as an ethical practice of care? For everyday use by publics outside of the healthcare system, the use of washable fabric masks is increasingly promoted as a sustainable practice of care for the environment. Detailed and easy-to-understand instructions relating to the wear, disposal of single-use masks and cleaning of cloth masks should be a key element of COVID education campaigns (UCL Plastic Waste Innovation Hub, 2020).

Some health authorities are beginning to provide advice to publics on how best to dispose of single-use masks. For example, the Australian statutory authority Sustainability Victoria (2020) offers guidelines on its website for what they describe as 'eco-friendly alternatives' to disposable masks. The site recommends cloth masks that are sourced locally from businesses that use ethical and sustainable work practices and materials and provides information about how to make one's own mask. It links to another Victorian Government website that offers free reusable cloth masks to agencies that support vulnerable people. Sustainability Victoria also details how plastic masks should be disposed: including cutting of the ear loops with scissors to protect wildlife before placing in the garbage. Similarly, organisations such as workplaces and educational institutions have become concerned that people are disposing inappropriately of their masks, including dropping them on the ground rather than placing them in garbage bins, or attempting to flush them down the toilet. For example, Figure 6.3 shows a sign in a Sydney university advising people how best to dispose of their masks.

As these guidelines suggest, the use of a COVID masks is a fraught practice of care that involves consideration of several ethical issues. It involves being

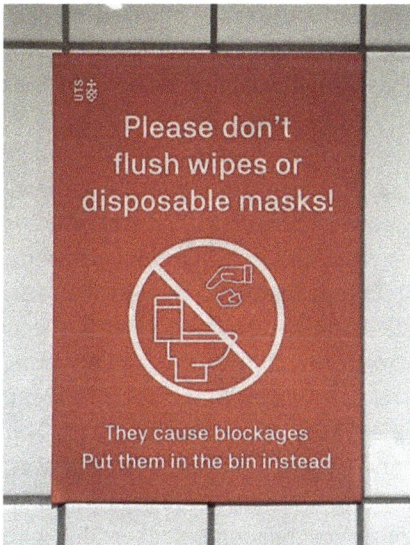

Figure 6.3: Sign in a Sydney university public toilet. Photo credit: Deborah Lupton.

careful through appropriate COVID mask purchasing, manufacture, distribution, laundering, disposal practices to both perform care effectively and extend that care to nonhuman others. This is an unusual perspective in COVID mask discourses, which have tended to focus exclusively on the health, wellbeing and rights of humans. Indeed, the burden placed on the environment through the production and disposal of masks both reflects and emerges through the prioritising of human life, health and wellbeing over that of the nonhuman. This deep-seated approach reflects historical and contemporary philosophies that value and recognise human exceptionalism and agency over nonhuman agencies. In this arrangement, the natural or more than human world is seen as passive and there to be shaped by the agentic human subject. Contemporary scholarship has articulated the constraints, and devastating impact of such human-centric thinking for environments and ecosystems, as well as human societies (Bennett, 2009; Alaimo, 2016; Haraway, 2016; Hernández et al., 2020).

In failing to recognise our entangled interdependence, conditions for both human and nonhuman entities are constrained, and our futures become increasingly precarious and uncertain. The COVID pandemic itself provides a compelling example (Chen, 2020; Braidotti, 2020). Human encroachment into wildlife habitats and the disruption of ecosystems by human industrial, agricultural and social practices, as well as actions such as trapping and consuming wild animals, have been linked to the emergence of zoonotic diseases (those that can jump between animals), of which COVID-19 is one (Mackenzie and Smith,

2020). In 2016, the United Nations published a report that connected human-driven damage to the environment to the emergence of pandemics (UNEP, 2016). Four years later, a pandemic that has already claimed millions of lives globally is running its course, with no sign yet of abating.

Feminist new materialism scholarship and Indigenous and First Nations cosmologies present alternatives to human-centred understandings of the environment. They have long clamoured for a dramatic reconfiguring of human relations with the more than human world. Such approaches, while not homogeneous, emphasise the entangled, co-constitutive relations that exist between humans, environments and non-human species, while accounting for the material effects of power relations. The notion of care is woven through this work, underscoring the need for a relational, responsive approach to care (Watts, 2013; TallBear, 2014; Bawaka Country et al., 2016; Todd, 2016; Puig de la Bellacasa, 2017; Hernández et al., 2020). Karen Barad and Donna Haraway (Barad, 2007; Haraway, 2013) acknowledge this need through the concept of 'response-ability', which involves learning to live with and take seriously our relational connections with human and nonhuman others. Bringing Indigenous and First Nations cosmologies to bear on care involves positioning ourselves as part of kin-networks so that we can work on 'kin-making' with nonhuman others (Hernández et al., 2020). When we notice and take seriously these intra-actions, we may loosen our grip on anthropocentrism and create new ways of thinking, knowing and being (Barad, 2007; TallBear, 2014; Alaimo, 2016; Todd, 2016; Hernández et al., 2020) that allow us to flourish in our more-than-human interconnectedness. Thinking through more-than-human perspectives, we are called upon to acknowledge and respond (ethically) to the broader sociopolitical and material conditions through which COVID masks and mask wearing emerge as meaningful (Braidotti, 2020). These respondings are always emergent and contingent.

Conclusion

In this chapter, we have discussed how face masks and the act of wearing (or not wearing) a mask are implicated in performative practices of care that emerge through the specific sociomaterial and political conditions of the COVID crisis. Face masks emerge as part of assemblages of care constituted through networks of humans, technologies, political and affective forces and nonhuman others. If we engage with more-than-human thinking when considering the phenomenon of the COVID mask, we recognise that although we don masks to protect ourselves and care for human others, masks continue to have sociomaterial effects long after they have been consumed for human purpose. The production of masks, the kind of masks we choose to wear, and how we dispose of them has implications for the more-than-human world. 'Thinking with care' (Puig de la Bellacasa, 2017), acting with 'response-ability' (Barad, 2007; Haraway, 2013) and 'kin-making' (Hernández et al., 2020) includes exposing and resisting our tendency to adopt a human-centric position in a post-COVID world in which our connections with and to nonhuman others have become more apparent and important than ever. Considering care from a more-than human-perspective requires responsiveness and carefulness so as not to resort to humanist tendencies that prioritise human experience when we work to imagine collective futures. As we imagine these futures, we will need to negotiate the tension between human-centred understandings of 'health' and the needs of the more-than-human world of which we are inextricably a part. As Geller (2020) reminds us, 'Wearing a washable, reusable mask shows that you care about other humans, as well as mammals, birds, fish, coral, and living organisms, in general'.

Epilogue: Masks Matter

Face masks have become central sociomaterial objects in the post-COVID world. Viewed from one perspective, they are nothing more than pieces of flimsy plastic or cloth fabric. From other perspectives, they bear intense symbolic weight and potentially threaten the health of the planet as well as acting as our only barrier between air-borne novel coronavirus infection. As Joe Biden announced in response to Donald Trump's dismissal of the importance of COVID masks: 'Masks matter'. Masks worn for health reasons have a complex cultural history, but the advent of the COVID crisis has both intensified and expanded the ways masks have come to matter as sociomaterial objects. In this book, we have surfaced the manifold complexities, forces and intra-actions of symbolism, discourse, politics, culture and embodiment in which the face mask in COVID is entangled. We have shown how, in the space of a year, face masks have materialised in many nooks and crannies of popular culture and quotidian life: from political protests to kitschy Christmas ornaments.

In a world in which the COVID crisis is still gathering momentum with little sign of slowing, great attention has been paid to initiatives of science and medical researchers in finding solutions to the crisis. While this research is undoubtedly vital, we have demonstrated in this book how important a social inquiry perspective is to complement it. Simple, time-honoured social practices such as hand washing, quarantine and physical distancing – and mask wearing – remain as integral to the COVID response globally as have the development of vaccines and new therapies for treating serious COVID disease. Surfacing and discussing the more-than-human environments in which COVID masking takes place – or is rejected or avoided – are key elements to future successful initiatives in fighting the spread of the novel coronavirus. Adopting this approach means a more-than-human body of research in health and medicine that can generate new, transversal and different ways of caring and better forms of governance and stewardship of more-than-human worlds and more-than-human health.

As we have argued, once we delve below the surface, the face mask is anything but a 'simple' preventive agent against infection with the coronavirus. The mask touches our skin, tugs at our ears, covers our facial expressions, fogs our spectacles, muffles our voices. Depending on what kind of materials it is made of and how many layers it has, a mask can be more or less difficult to breathe with. Masks are hard to wear for too long. They get uncomfortable, damp and soiled. Attention must be paid to how we deal with our masks. We must be trained to handle them carefully, dispose of them thoughtfully and to care for reusable

https://doi.org/10.1515/9783110723717-008

masks by storing and washing them appropriately so that they remain an effective hygienic item that prevents rather than promotes coronavirus spread. Masks can feel uncomfortable, but they can also feel good. They can make us feel safe and that we are demonstrating our moral agency as responsible citizens who care about others. Hand-crafted masks are made with and for care, as symbols of love and recognition of need, of creativity and making cultures. Mass-produced disposable masks are also made by people who are coerced into slave labour to meet the demands of wealthy countries. Masks can be discarded carelessly, treated with disdain, flung down after wear and accumulating to pollute the planet and thereby becoming 'matter out of place'. For those who rail against mask wearing, masks are viewed as constricting of the body and of political beliefs, rejected out of hand.

Although we may not all live and experience the COVID crisis in the same ways, we are all implicated. The same social and political systems that constrain conditions of possibility for some people are those that produce conditions conducive to health and wellbeing for others. As Rosi Braidotti (2020) has written: 'We are in this together, but we are not one and the same'. We have identified and explained *how* and *why* masks matter: not just because they can save lives, but in many other ways as well. We have feelings about and feelings with masks: they bring the sensory together with the affective. Masks protect us from nonhuman entities such as the novel coronavirus, but simultaneously have sociomaterial effects on the environment and ecosystem. They are things that come together with human bodies, other living and non-living things, place and space to generate forms of safety, self-expression, embodied socialities, creativity and care but also disturbing affective forces such as anger, frustration, discomfort, racism, social exclusion and shame. This is why masks as sociomaterial objects in COVID times are so vibrant, so political, so meaningful, so forceful, so enchanting: in short, so full of thing-power.

References

ABC News (2020) Authorities remove almost a million N95 masks and other supplies from alleged hoarder. Available at: https://www.youtube.com/watch?v=MmNqXaGuo2k&ref= hvper.com&utm_source=hvper.com&utm_medium=website

Adyel TM (2020) Accumulation of plastic waste during COVID-19. *Science* 369(6509): 1314.

Agar J (2013) *Constant Touch.* London: Icon Books.

Alaimo S (2010) *Bodily Natures: Science, Environment, and the Material Self.* Bloomington: Indiana University Press.

Alaimo S (2016) *Exposed: Environmental Politics and Pleasures in Posthuman Times.* Minneapolis: University of Minnesota Press.

Alaimo S (2018) Trans-corporeality. In: Braidotti R and Hlavajova M (eds) *Posthuman Glossary.* London: Bloomsbury, pp. 435–438.

Allen IK (2020) Thinking with a feminist political ecology of air-and-breathing-bodies. *Body & Society* 26(2): 79–105.

American Academy of Dermatology Association (2020) Protecting your skin while using PPE. Available at: https://www.aad.org/member/practice/coronavirus/clinical-guidance/ppe-medical-supplies

American Hospital Association (2020) 100 million mask challenge. Available at: https://www.100millionmasks.org/

Amir-Moazami S (2016) The secular embodiments of face-veil controversies across Europe. In: Göle N (ed) *Islam and Public Controversy in Europe.* London: Routledge, pp. 83–100.

Anadolu Agency (2020) Nigeria: COVID-19 patients protest over 'ill treatment'. *Anadolu Agency.* Available at: https://www.aa.com.tr/en/africa/nigeria-covid-19-patients-protest-over-ill-treatment/1830966

Andrews GJ and Duff C (2019) Matter beginning to matter: on posthumanist understandings of the vital emergence of health. *Social Science & Medicine* 226: 123–134.

Anguyo I (2020) Unmasking political COVID-19 face coverings in Uganda. *Africa at LSE.* Available at: http://eprints.lse.ac.uk/106309/

Anonymous (2020a) Which countries have made wearing face masks compulsory? *Aljazeera.* Available at: https://www.aljazeera.com/news/2020/08/17/which-countries-have-made-wearing-face-masks-compulsory/

Anonymous (2020b) Japan to give two masks each to 50 million households to fight virus. *The Japan Times.* Available at: https://www.japantimes.co.jp/news/2020/04/02/national/japanese-government-distribute-two-masks-per-household-abe/

Anonymous (2020c) Israel's new virus surveillance system detects masks and social distancing; no facial recognition capability. *Security Magazine.* Available at: https://www.securitymagazine.com/articles/93891-israels-new-virus-surveillance-system-detects-masks-and-social-distancing-no-facial-recognition-capability

Apata GO (2020) 'I Can't Breathe': The suffocating nature of racism. *Theory, Culture & Society* 37(7–8): 241–254.

APM Research Lab (2020) COVID-19 deaths analyzed by race and ethnicity. Available at: https://www.apmresearchlab.org/covid/deaths-by-race

Aragaw TA (2020) Surgical face masks as a potential source for microplastic pollution in the COVID-19 scenario. *Marine Pollution Bulletin*, 159. Available at: http://www.sciencedirect.com/science/article/pii/S0025326X20306354

https://doi.org/10.1515/9783110723717-009

Associated Press (2020) Refugees protest under coronavirus lockdown in Rwanda. *VOA News.* Available at: https://www.voanews.com/covid-19-pandemic/refugees-protest-under-coro navirus-lockdown-rwanda

Australian Bureau of Statistics (2020) Household impacts of COVID-19 survey, October 2020. Available at: https://www.abs.gov.au/statistics/people/people-and-communities/house hold-impacts-covid-19-survey/sep-2020#precautions

Australian Government Department of Health (2020a) Coronavirus (COVID-19) – How to make a cloth mask. Available at: https://www.health.gov.au/resources/publications/how-to-make-cloth-mask

Australian Government Department of Health (2020b) Should I wear a face mask in public? Available at: https://www.health.gov.au/news/should-i-wear-a-face-mask-in-public-0

Bambra C, Riordan R, Ford J, et al. (2020) The COVID-19 pandemic and health inequalities. *Journal of Epidemiology and Community Health* 74: 964–968.

Barad K (2003) Posthumanist performativity: Toward an understanding of how matter comes to matter. *Signs* 28(3): 801–831.

Barad K (2007) *Meeting the Universe Halfway: Quantum Physics and the Entanglement of Matter and Meaning.* Durham: Duke University Press.

Bashford A (2016) (ed.) *Quarantine: Local and Global Histories.* Houndmills: Palgrave Macmillan.

Bawaka Country, Wright S, Suchet-Pearson S, et al. (2016) Co-becoming Bawaka: towards a relational understanding of place/space. *Progress in Human Geography* 40(4): 455–475.

Beaudry MC (2006) *Findings: The Material Culture of Needlework and Sewing.* New Haven: Yale University Press.

Bennett J (2001) *The Enchantment of Modern Life: Attachments, Crossings, and Ethics.* Princeton: Princeton University Press.

Bennett J (2004) The force of things: steps toward an ecology of matter. *Political Theory* 32(3): 347–372.

Bennett J (2009) *Vibrant Matter: A Political Ecology of Things.* Durham: Duke University Press.

Beresford J (2020) Anti-mask demonstrators stage sit-in protest in Dublin. *The Irish Post.* Available at: https://www.irishpost.com/news/anti-mask-demonstrators-stage-sit-in-pro test-in-dublin-194384

Berger JM (2016) Without prejudice: what sovereign citizens believe. Available at: https://ex tremism.gwu.edu/sites/g/files/zaxdzs2191/f/downloads/JMB%20Sovereign%20Citizens. pdf

Berker T, Hartmann M, Punie Y, et al. (2006) Introduction. In: Berker T, Hartmann M, Punie Y, et al. (eds) *Domestication of Media and Technology.* Berkshire: Open University Press, pp. 1–16.

Bhasin T, Butcher C, Gordon E, et al. (2020) Does Karen wear a mask? The gendering of COVID-19 masking rhetoric. *International Journal of Social Policy* 40(9/10): 929–937.

Borresen K (2020) How to store face masks to ensure they're safe and clean. *Huffington Post.* Available at: https://www.huffingtonpost.ca/entry/storing-face-mask-safely_ca_ 5f5a5515c5b62874bc19d73e

Bracken RC (2020) Influenza and embodied sociality in early twentieth-century American literature. *American Literary History* 32(3): 507–534.

Bradley J (2020) In scramble for coronavirus supplies, rich countries push poor aside. Available at: https://www.nytimes.com/2020/04/09/world/coronavirus-equipment-rich-poor.html

Bradsher K and Swanson A (2020) The U.S. needs China's masks, as acrimony grows. *The New York Times*. Available at: https://www.nytimes.com/2020/03/23/business/coronavi rus-china-masks.html

Braidotti R (2008) Of poststructuralist ethics and nomadic subjects. In: Düwell M, Rehmann-Sutter C and Mieth D (eds) *The Contingent Nature of Life: Bioethics and Limits of Human Existence*. Dordrecht: Springer Netherlands, pp. 25–36.

Braidotti R (2019) *Posthuman Knowledge*. Cambridge: Polity.

Braidotti R (2020) 'We' are in this together, but we are not one and the same. *Journal of Bioethical Inquiry* 17: 465–469.

Brenan M (2020) More mask use, worry about lack of social distancing in U.S. *Gallup*. Available at: https://news.gallup.com/poll/313463/mask-worry-lack-social-distancing. aspx

Brickell S (2020) These face brackets for masks will make breathing so much easier. *Health.com*. Available at: https://www.health.com/condition/infectious-diseases/coronavi rus/face-bracket-for-mask

Burfoot A (2010) Feminist technoscience: a solution to the theoretical conundrums and the wave of feminist politics? *Resources for Feminist Research* 33(3/4): 71–93.

Burgess A and Horii M (2012) Risk, ritual and health responsibilisation: Japan's 'safety blanket'of surgical face mask-wearing. *Sociology of Health & Illness* 34(8): 1184–1198.

Burgess J and Green J (2018) *YouTube: Online Video and Participatory Culture*. Cambridge: Polity.

Buse C, Martin D and Nettleton S (2018) Conceptualising 'materialities of care': making visible mundane material culture in health and social care contexts. *Sociology of Health & Illness* 40(2): 243–255.

Calvo RA, Deterding S and Ryan RM (2020) Health surveillance during covid-19 pandemic. *BMJ* 369. Available at: https://www.bmj.com/content/369/bmj.m1373

Capraro V and Barcelo H (2020) The effect of messaging and gender on intentions to wear a face covering to slow down COVID-19 transmission. *PsyArXiv Preprints*. Available at: https://psyarxiv.com/tg7vz

Centers for Disease Control and Prevention (2020a) Considerations for wearing masks. Available at: https://www.cdc.gov/coronavirus/2019-ncov/prevent-getting-sick/cloth-face-cover-guidance.html

Centers for Disease Control and Prevention (2020b) Scientific brief: SARS-CoV-2 and potential airborne transmission. Available at: https://www.cdc.gov/coronavirus/2019-ncov/more/ scientific-brief-sars-cov-2.html

Centers for Disease Control and Prevention (2020c) Scientific brief: community use of cloth masks to control the spread of SARS-CoV-2. Available at: https://www.cdc.gov/coronavi rus/2019-ncov/more/masking-science-sars-cov2.html

Centers for Disease Control and Prevention (2020d) How to make masks. Available at: https://www.cdc.gov/coronavirus/2019-ncov/prevent-getting-sick/how-to-make-cloth-face-covering.html

Chen MY (2020) Feminisms in the air. *Signs*. Available at http://signsjournal.org/covid/chen/

Cheng KK, Lam TH and Leung CC (2020) Wearing face masks in the community during the COVID-19 pandemic: altruism and solidarity. *The Lancet.* Available at: https://www.the lancet.com/journals/lancet/article/PIIS0140-6736(20)30918-1/fulltext

Cheung Y (2020) Enter the 'Quaranzine': zines that boost resistance, mutual aid, and self-care. *Hyperallergic.* Available at: https://hyperallergic.com/560443/enter-the-quaranzine-zines-that-boost-resistance-mutual-aid-and-self-care/

Coates H (2020) These are the face masks that don't cause 'maskne'. *Vogue.* Available at: https://www.vogue.co.uk/beauty/article/silk-face-masks-maskne

Cole L (2010) Of mice and moisture: rats, witches, miasma, and early modern theories of contagion. *Journal for Early Modern Cultural Studies* 10(2): 65–84.

Couch DL, Robinson P and Komesaroff PA (2020) COVID-19 – extending surveillance and the panopticon. *Journal of Bioethical Inquiry* online first.

Crăciun M (2017) Aesthetics, ethics and fashionable veiling: a debate in contemporary Turkey. *World Art* 7(2): 329–352.

Cramer M and Sheikh K (2020) Surgeon General urges the public to stop buying face masks. *The New York Times.* Available at: https://www.nytimes.com/2020/02/29/health/corona virus-n95-face-masks.html

Davidson H (2020a) Hong Kong face masks ban largely upheld despite coronavirus. *The Guardian.* Available at: http://www.theguardian.com/world/2020/apr/09/hong-kong-court-upholds-face-masks-ban-despite-coronavirus

Davidson H (2020b) The story of David Dungay and an Indigenous death in custody. *The Guardian.* Available at: http://www.theguardian.com/australia-news/2020/jun/11/the-story-of-david-dungay-and-an-indigenous-death-in-custod

Department of Health and Human Services Victoria (2020) Face coverings: whole of Victoria. Available at: https://www.dhhs.vic.gov.au/face-coverings-covid-19

Dicker R (2020) Tomi Lahren ticks off Twitter with 'purse' dig at Biden's mask-wearing. *Huffington Post.* Available at: https://www.huffpost.com/entry/tomi-lahren-twitter-joe-biden-purse_n_5f7c7454c5b6e5aba0d0534c

Dolan B (2020) Unmasking history: who was behind the Anti-Mask League protests during the 1918 influenza epidemic in San Francisco? *Perspectives in Medical Humanities* 5(19): 1–23.

Dolphijn R and Van der Tuin I (2012) *New Materialism: Interviews & Cartographies.* Open Humanities Press. Available at http://www.openhumanitiespress.org/books/titles/new-materialism/#:~:text=edited%20by%20Rick%20Dolphijn%20and,1%2D60785%2D281%2D0

Domingues da Silva DB (2020) Perspective | 'I can't breathe' is tied to a long history of Black asphyxiation. *The Washington Post.* Available at: https://www.washingtonpost.com/out look/2020/08/14/i-cant-breathe-is-tied-long-history-black-asphyxiation/

Dove (2020) Take care be safe: how we're caring for our community. Available at: https://www.dove.com/us/en/stories/about-dove/take-care-be-safe.html

Easterbrook-Smith G (2020) By bread alone: baking as leisure, performance, sustenance, during the COVID-19 crisis. *Leisure Sciences* online first.

Etsy (2020) Tips for selling handmade masks and face covers on Etsy. Available at: https://www.etsy.com/au/seller-handbook/article/tips-for-selling-handmade-masks-and-face/788845527234

Factcheck AFP (2020) World Health Organization refutes viral claims that holding your breath can test for COVID-19. Available at: https://factcheck.afp.com/world-health-organization-refutes-viral-claims-holding-your-breath-can-test-covid-19

Fadare OO and Okoffo ED (2020) Covid-19 face masks: A potential source of microplastic fibers in the environment. *Science of the Total Environment* 737. Available at https://www.sciencedirect.com/science/article/abs/pii/S0048969720338006?via%3Dihub

Fine M and Tronto J (2020) Care goes viral: care theory and research confront the global COVID-19 pandemic. *International Journal of Care and Caring* 4(3): 301–309.

Fischer EP, Fischer MC, Grass D, et al. (2020) Low-cost measurement of face mask efficacy for filtering expelled droplets during speech. *Science Advances* 6(36): 1–5.

Flaskerud JH (2020) Masks, politics, culture and health. *Issues in Mental Health Nursing* 41(9): 846–849.

Fleisher O, Giandordoli G, Parshina-Kottas Y, et al. (2020) Masks work. Really. We'll show you how. *The New York Times*. Available at: https://www.nytimes.com/interactive/2020/10/30/science/wear-mask-covid-particles-ul.html

Foucault M (1965) *Madness and Civilization: the History of Insanity in the Age of Reason*. New York: Random House.

Foucault M (1975) *The Birth of the Clinic: An Archaeology of Medical Perception*. New York: Vintage Books.

Foucault M (1977) *Discipline and Punish: the Birth of the Prison*. London: Allen Lane.

Foucault M (1984) The politics of health in the eighteenth century. *The Foucault Reader*. New York: Pantheon Books, pp. 273–289.

Foucault M (1986) *The Care of the Self: The History of Sexuality Volume 3*. New York: Pantheon.

Fox NJ and Alldred P (2017) *Sociology and the New Materialism: Theory, Research, Action*. London: Sage.

Freund T (2020) There exists a sheet mask you can wear under your face mask. *Marie Claire*. Available at: https://www.marieclaire.com/beauty/a34114118/sheet-mask-under-face-mask/

Friedman G (2020) McDonald's joins Walmart and dozens of other chains with mask mandates. *The New York Times*. Available at: https://www.nytimes.com/article/which-stores-require-masks.html

Gammon S and Ramshaw G (2020) Distancing from the present: nostalgia and leisure in lockdown. *Leisure Sciences* online first.

Geller PL (2020) The archeology of the disposable face mask. *Slate*. Available at: https://slate.com/technology/2020/10/disposable-masks-ocean-pollution-archaeology.html

Gereffi G (2020) What does the COVID-19 pandemic teach us about global value chains? The case of medical supplies. *Journal of International Business Policy* 3(3): 287–301.

Gharib M and Harlan B (2020) #Quaranzine round-up: a look at pandemic life through the pages of your mini-zines. *NPR*. Available at: https://www.npr.org/2020/07/18/890809921/-quaranzine-round-up-a-look-at-pandemic-life-through-the-pages-of-your-mini-zine

Gillespie E (2020) The rise of 'sovereign people' and why they argue laws don't apply to them. *SBS News*. Available at: https://www.sbs.com.au/news/the-feed/the-rise-of-sovereign-people-and-why-they-argue-laws-don-t-apply-to-them

Górska M (2016) *Breathing Matters: Feminist Intersectional Politics of Vulnerability.* Linköping: Linköping University Electronic Press.

Greenhalgh T, Schmid MB, Czypionka T, et al. (2020) Face masks for the public during the covid-19 crisis. *British Medical Journal,* 369. Available at: http://www.bmj.com/content/369/bmj.m1435.abstract

Greenwood F (2020) The dawn of the shout drone. *Slate.* Available at: https://slate.com/technology/2020/04/coronavirus-shout-drone-police-surveillance.html

Grosz E (1994) *Volatile Bodies: Toward a Corporeal Feminism.* Sydney: Allen & Unwin.

Grote H and Izagaren F (2020) Covid-19: The communication needs of D/deaf healthcare workers and patients are being forgotten. *BMJ* 369. Available at https://www.bmj.com/content/369/bmj.m2372.full

Guardian staff and agencies (2020) 'It isn't Mad Max': women charged after fight over toilet paper in Sydney. *The Guardian.* Available at: http://www.theguardian.com/australia-news/2020/mar/07/it-isnt-mad-max-police-warning-after-shoppers-brawl-over-toilet-paper-in-sydney

Gunders P (2020) What to look for when buying reusable face masks in Australia and how you need to clean them. *ABC News.* Available at: https://www.abc.net.au/news/2020-07-20/coronavirus-reusable-face-mask-guide-melbourne-what-to-look-for/12471918

Haggerty M (2020) The homemade masks of coronavirus. *Vox.* Available at: https://www.vox.com/the-goods/2020/4/16/21222370/homemade-mask-photos-coronavirus-ppe

Han C, Shi J, Chen Y, et al. (2020) Increased flare of acne caused by long-time mask wearing during COVID-19 pandemic among general population. *Dermatological Therapy* 33(4): 1–3.

Haraway D (1988) Situated knowledges: the science question in feminism and the privilege of partial perspective. *Feminist Studies* 14(3): 575–599.

Haraway D (2013) *When Species Meet.* Minneapolis: University of Minnesota Press.

Haraway D (2016) *Staying with the Trouble: Making Kin in the Chthulucene.* Durham: Duke University Press.

Harris K (2020) Canadians should wear masks as 'an added layer of protection,' says Tam. *CBC.* Available at: https://www.cbc.ca/news/politics/masks-covid-19-pandemic-public-health-1.5576895

Hernández JC (2020) In China, where the coronavirus pandemic began, life is starting to look normal. *The New York Times.* Available at https://www.nytimes.com/2020/08/23/world/asia/china-coronavirus-normal-life.html

Hernández K, Rubis JM, Theriault N, et al. (2020) The Creatures Collective: manifestings. *Environment and Planning E: Nature and Space* online first.

Horii M (2014) Why do the Japanese wear masks? *Electronic Journal of Contemporary Japanese Studies,* 14. Available at: http://www.japanesestudies.org.uk/ejcjs/vol14/iss2/horii.html

Horvath H (2020) How to shop for face masks on Etsy, according to experts. *NBC News.* Available at: https://www.nbcnews.com/shopping/apparel/best-face-masks-etsy-n1228621

Howard J (2020) To help stop coronavirus, everyone should be wearing face masks. The science is clear. *The Guardian.* Available at: http://www.theguardian.com/commentisfree/2020/apr/04/why-wear-a-mask-may-be-our-best-weapon-to-stop-coronavirus

Irigaray L (1999) *The Forgetting of Air in Martin Heidegger.* Austin: University of Texas Press.

Irigaray L (2002) *Between East and West: From Singularity to Community.* New York: Columbia University Press.

Jackson AY and Mazzei LA (2012) *Thinking with Theory in Qualitative Research.* New York: Taylor & Francis.

Jamieson L (2011) Intimacy as a concept: explaining social change in the context of globalisation or another form of ethnocentricism? *Sociological Research Online* 16(4). Available at https://journals.sagepub.com/doi/full/10.5153/sro.2497

Janzwood S and Lee M (2020) Behind the mask: anti-mask and pro-mask attitudes in North America. *Cascade Institute.* Brief no. 6. Available at https://cascadeinstitute.org/isc-brief/behind-the-mask-anti-mask-and-pro-mask-attitudes-in-north-america/

Javid B, Weekes MP and Matheson NJ (2020) Covid-19: should the public wear face masks? *British Medical Journal*, 369. Available at: http://www.bmj.com/content/369/bmj.m1442.abstract

Kassam A (2020) 'More masks than jellyfish': coronavirus waste ends up in the ocean. *The Guardian.* Available at: https://www.theguardian.com/environment/2020/jun/08/more-masks-than-jellyfish-coronavirus-waste-ends-up-in-ocean

Kavilanz P (2020) Sudden sewing boom has sewing machine sellers scrambling. *CNN Business.* Available at: https://edition.cnn.com/2020/08/13/business/sewing-machines-demand/index.html

Kayat S (2020) Doctor's Note: Do masks protect us from coronavirus? *Aljazeera.* Available at: https://www.aljazeera.com/features/2020/04/19/doctors-note-do-masks-protect-us-from-coronavirus/

Kelleher SR (2020) Neck gaiters do not curb COVID-19, study finds. *Forbes Magazine.* Available at: https://www.forbes.com/sites/suzannerowankelleher/2020/08/11/neck-gaiters-do-not-curb-covid-19-study-finds/

Kelley D (2020) The person within the mask: mask-wearing, identity, and communication. *American Journal of Qualitative Research* 4(3): 111–130.

Koay A (2020) 'I am a Sovereign. I am "We the People"': Rant by maskless S'porean woman in Shunfu, explained. *Mothership.* Available at: https://mothership.sg/2020/05/sovereign/

Kourlas G (2020) How we use our bodies to navigate a pandemic. *New York Times.* Available at: https://www.nytimes.com/2020/03/31/arts/dance/choreographing-the-street-coronavirus.html

Kwon D (2020) Infographic: What we know about how masks can slow disease spread. *The Scientist.* Available at: https://www.the-scientist.com/infographics/infographic-what-we-know-about-how-masks-can-slow-disease-spread-67712

Leach TB (2020) How to sew a quick and easy cloth face mask: with medical-grade masks in short supply, try making a DIY alternative. *AARP.* Available at: https://www.aarp.org/health/healthy-living/info-2020/making-cloth-face-masks.html

Lehrer R (2020) The virus has stolen your face from me. *The New York Times.* Available at: https://www.nytimes.com/2020/12/10/opinion/coronavirus-mask-faces-art.html?auth=linked-google&campaign_id=39&emc=edit_ty_20201210&instance_id=24893&nl=opinion-today&regi_id=135924415&segment_id=46553&te=1&user_id=ed8e99580e16b8dd577c628c9a49de1e

Lenton P, Shaw R, Earp J, et al. (2020) The internet is showing support for the workers who calmly dealt With 'Bunnings Karen'. *Junkee*. Available at: https://junkee.com/bunnings-karen-meme/263574

Levinas E (1979) *Totality and Infinity*. Boston: Martinus Nijhoff Publishers.

Li C and McElveen R (2020) Mask diplomacy: coronavirus upended generations of China-Japan antagonism. *China-US Focus*. Available at: https://www.chinausfocus.com/foreign-policy/mask-diplomacy-how-coronavirus-upended-generations-of-china-japan-antagonism

Life of Breath (2020) Available at: https://lifeofbreath.org/

Lizza R and Lippman D (2020) Wearing a mask is for smug liberals. Refusing to is for reckless Republicans. *Politico*. Available at: https://www.politico.com/news/2020/05/01/masks-politics-coronavirus-227765

Lockwood C and Jordan Z (2020) 13 insider tips on how to wear a mask without your glasses fogging up, getting short of breath or your ears hurting. *The Conversation*. Available at: http://theconversation.com/13-insider-tips-on-how-to-wear-a-mask-without-your-glasses-fogging-up-getting-short-of-breath-or-your-ears-hurting-143001

Luckman S (2013) The aura of the analogue in a digital age: women's crafts, creative markets and home-based labour after Etsy. *Cultural Studies Review* 19(1): 249–270.

Luckman S (2015) *Craft and the Creative Economy*. Hampshire: Palgrave MacMillan.

Lupton D (1995) *The Imperative of Health: Public Health and the Regulated Body*. London: Sage.

Lupton D (2012) *Medicine as Culture: Illness, Disease and the Body*. London: Sage.

Lupton D (2015) *Digital Sociology*. London: Routledge.

Lupton D (2019) Toward a more-than-human analysis of digital health: inspirations from feminist new materialism. *Qualitative Health Research* 29(14): 1998–2006.

Lupton D (2020a) A more-than-human approach to bioethics: The example of digital health. *Bioethics* 34(9): 969–976.

Lupton D (2020b) Special section on 'Sociology and the Coronavirus (COVID-19) Pandemic'. *Health Sociology Review* 29(2): 111–112.

Lupton D (2021) Contextualising COVID-19: sociocultural perspectives on contagion. In: Lupton D and Willis K (eds) *The COVID-19 Crisis: Social Perspectives*. London: Routledge, pp. 14–24.

Lupton D and Willis K (2021) COVID Society: introduction to the book. In: Lupton D and Willis K (eds) *The COVID-19 Crisis: Social Perspectives*. London: Routledge, pp. 3–13.

Lynteris C (2018) Plague masks: the visual emergence of anti-epidemic personal protection equipment. *Medical Anthropology* 37(6): 442–457.

Lyon D (2018) *The Culture of Surveillance: Watching as a Way of Life*. New York: John Wiley & Sons.

Ma Y and Zhan N (2020) To mask or not to mask amid the COVID-19 pandemic: how Chinese students in America experience and cope with stigma. *Chinese Sociological Review* online first.

Mackenzie JS and Smith DW (2020) COVID-19: a novel zoonotic disease caused by a coronavirus from China: what we know and what we don't. *Microbiology Australia* online first.

Mai X, Ge Y, Tao L, et al. (2011) Eyes are windows to the Chinese soul: evidence from the detection of real and fake smiles. *PLoS One* 6(5). Available at: https://journals.plos.org/plosone/article?id=10.1371/journal.pone.0019903

Maker S (2020) 5 ways to make a no-sew face mask with household materials. Available at: https://sarahmaker.com/how-to-make-a-no-sew-face-mask-with-at-home-materials/

Malabou C (2008) Addiction and grace: preface to Félix Ravaisson's Of Habit. In: Carlisle C and Sinclair M (eds) *Félix Ravaisson: Of Habit*. London, UK: Continuum Books.

Malpass A, Dodd J, Feder G, et al. (2019) Disrupted breath, songlines of breathlessness: an interdisciplinary response. *Medical Humanities* 45(3): 294–303.

Margit M (2020) Now in use: Israeli facial-recognition system that can 'see through' masks. *The Media Line*. Available at: https://themedialine.org/life-lines/now-in-use-israeli-facial-recognition-system-that-can-see-through-masks-video-report/

Marres N (2016) *Material Participation: Technology, the Environment and Everyday Publics*. Houndmills: Palgrave MacMillan.

#Masks4All (2020a) *Masks4All*. Available at: https://masks4all.co/

#Masks4All (2020b) What countries require masks in public or recommend masks? Available at: https://masks4all.co/what-countries-require-masks-in-public/

Martin A, Myers N and Viseu A (2015) The politics of care in technoscience. *Social Studies of Science* 45(5): 625–641.

Martin GP, Hanna E, McCartney M, et al. (2020) Science, society, and policy in the face of uncertainty: reflections on the debate around face coverings for the public during COVID-19. *Critical Public Health* 30(5): 501–508.

Mathers M (2020) Coronavirus: Chinese firms 'using forced Uighur labour to produce face masks' for US. *The Independent*. Available at: https://www.independent.co.uk/news/world/asia/coronavirus-chinese-companies-uighur-labour-face-masks-ppe-a9628836.html

Mathew S (2020) Coronavirus: the latecomer's guide to face masks. *The Hindu*. Available at: https://www.thehindu.com/news/cities/Delhi/the-latecomers-guide-to-face-masks/article32312115.ece

Matuschek C, Moll F, Fangerau H, et al. (2020) The history and value of face masks. *European Journal of Medical Research* 25(1). Available at https://eurjmedres.biomedcentral.com/articles/10.1186/s40001-020-00423-4

Mayo Clinic Health System (2020) Debunked myths about face masks. Available at: https://www.mayoclinichealthsystem.org/hometown-health/speaking-of-health/debunked-myths-about-face-masks

McDougall A, Kinsella EA, Goldszmidt M, et al. (2018) Beyond the realist turn: a socio-material analysis of heart failure self-care. *Sociology of Health & Illness* 40(1): 218–233.

McMahon DE, Peters GA, Ivers LC, et al. (2020) Global resource shortages during COVID-19: bad news for low-income countries. *PLOS Neglected Tropical Diseases* 14(7). Available at: https://journals.plos.org/plosntds/article?id=10.1371/journal.pntd.0008412

Metz N (2020) Commentary: Many of us are trying to figure out how to deal with this new normal. Maybe part of that process is acknowledging the yuck factor of masks. *The Chicago Tribune*. Available at: https://www.chicagotribune.com/coronavirus/ct-coronavirus-the-misery-of-masks-0512-20200512-nz3nuv7uyngl7mczdmz2hjll2e-story.html

Mills J (2020) Trouble breathing: 'We all breathe the same air, but we don't breathe equally'. *The Guardian*. Available at: http://www.theguardian.com/australia-news/2020/oct/04/trouble-breathing-we-all-breathe-the-same-air-but-we-dont-breathe-equally

Morrow G and Compagni G (2020) Mask mandates, misinformation, and data voids in local news coverage of COVID-19. *APSA Preprints*. Available at: http://doi.org/10.33774/apsa-2020-f3rnf

O'Kane S (2020) Amazon stops selling N95 and surgical masks to public. *The Verge*. Available at: https://www.theverge.com/2020/4/2/21204625/amazon-sales-ban-n95-public-surgical-hospitals-governments

OECD (2020a) The face mask global value chain in the COVID-19 outbreak: evidence and policy lessons. Available at: http://www.oecd.org/coronavirus/policy-responses/the-face-mask-global-value-chain-in-the-covid-19-outbreak-evidence-and-policy-lessons-a4df866d/

OECD (2020b) Trade interdependencies in COVID-19 goods. Available at: https://www.oecd.org/coronavirus/policy-responses/trade-interdependencies-in-covid-19-goods-79aaa1d6/

Oxley R and Russell A (2020) Interdisciplinary perspectives on breath, body and world. *Body & Society* 26(2): 3–29.

Palmer CL and Peterson RD (2020) Toxic Mask-ulinity: the link between masculine toughness and affective reactions to mask wearing in the COVID-19 era. *Politics & Gender* online first.

Palmer E (2020) Asian woman allegedly attacked in New York subway station for wearing protective mask. *Newsweek*. Available at: https://www.newsweek.com/new-york-subway-attack-coronavirus-woman-mask-1485842

Patty A (2020) Hospitals urge retailers to stop selling high-grade masks to public. *The Sydney Morning Herald*. Available at: https://www.smh.com.au/national/hospitals-urge-retailers-to-stop-selling-high-grade-masks-to-public-20200810-p55kbh.html

Pelling M (1993) Contagion/germ theory/specificity. *Companion Encyclopedia of the History of Medicine* 1: 309–334.

Pendo E, Gatter R and Mohapatra S (2020) Resolving tensions between disability rights law and COVID-19 mask policies. *Maryland Law Review Online* 80. Available at: https://digitalcommons.law.umaryland.edu/endnotes/68

Petersen A and Lupton D (1996) *The New Public Health: Health and Self in the Age of Risk*. London: Sage.

Philipose R (2020) Covid-19: A look at anti-mask rallies held around the world amid the pandemic. *The Indian Express*. Available at: https://indianexpress.com/article/world/covid-19-a-look-at-anti-mask-rallies-held-around-the-world-amid-the-pandemic-6585722/

Prata JC, Silva ALP, Walker TR, et al. (2020) COVID-19 pandemic repercussions on the use and management of plastics. *Environmental Science & Technology* 54(13): 7760–7765.

Price L and Hawkins H (2018) *Geographies of Making, Craft and Creativity*. London: Routledge.

Providence (2020) 100 Million Mask Challenge. Available at: https://www.providence.org/lp/100m-masks

Public Health England (2020) How to make a cloth face covering. Available at: https://www.gov.uk/government/publications/how-to-wear-and-make-a-cloth-face-covering/how-to-wear-and-make-a-cloth-face-covering

Puig de la Bellacasa M (2012) 'Nothing comes without its world': thinking with care. *The Sociological Review* 60(2): 197–216.

Puig de la Bellacasa M (2017) *Matters of Care: Speculative Ethics in More Than Human Worlds*. Minneapolis: University of Minnesota Press.

Puranam E (2020) India's migrant workers protest against lockdown extension. Available at: https://www.aljazeera.com/videos/2020/04/15/indias-migrant-workers-protest-against-lockdown-extension/

Randone A and Spencer K (2020) Where to find a fashion-forward face mask to express your personal style. *Refinery29*. Available at: https://www.refinery29.com/en-us/2020/05/9782883/stylish-fashion-face-mask-coronavirus

Ratto M and Boler M (2014) *DIY Citizenship: Critical Making and Social Media*. Massachusetts: MIT Press.

Readfearn G (2020) How ventilators work and why they are so important in saving people with coronavirus. *The Guardian*. Available at: http://www.theguardian.com/world/2020/mar/27/how-ventilators-work-and-why-they-are-so-important-in-saving-people-with-coronavirus

Red Cross Australia (2020) Wearing face masks. Available at: https://www.redcross.org.au/stories/covid-19/wearing-face-masks

Renold E (2018) 'Feel what I feel': making da(r)ta with teen girls for creative activisms on how sexual violence matters. *Journal of Gender Studies* 27(1): 37–55.

Ridler F (2020) Woman, 35, who stockpiled £2,500 of masks, gloves and face shields refuses to donate any to the NHS as they 'should have been more prepared'. *The Daily Mail*. Available at: https://www.dailymail.co.uk/news/article-8266591/Woman-35-stockpiled-2-500-PPE-refuses-donate-NHS.html

Roberts KP, Bowyer C, Kolstoe S, et al. (2020) Coronavirus face masks: An environmental disaster that might last generations. *The Conversation*. Available at: https://theconversation.com/coronavirus-face-masks-an-environmental-disaster-that-might-last-generations-144328

Rubin C (2020) Maskne is the new acne, and here's what is causing it. *The New York Times*. Available at: https://www.nytimes.com/article/maskne-acne.html

Ruiz NG, Horowitz JM and Tamir C (2020) Many Black and Asian American say that they have experienced discrimination amid the COVID-19 outbreak. Available at: https://www.pewsocialtrends.org/2020/07/01/many-black-and-asian-americans-say-they-have-experienced-discrimination-amid-the-covid-19-outbreak/

Sarteschi CM (2020) Sovereign citizens: a narrative review with implications of violence towards law enforcement. *Aggression and Violent Behavior* online first.

Savin J (2020) This is why you need an extra face mask on a rainy day. *Cosmopolitan*. Available at: https://www.cosmopolitan.com/uk/body/health/a34310929/face-mask-in-rain-effectiveness/

Schive K (2020) Are neck gaiters worse than no mask at all? Available at: https://medical.mit.edu/covid-19-updates/2020/08/neck-gaiters

Seale H (2020) It's easy to judge. But some people really can't wear a mask. *The Conversation*. Available at: http://theconversation.com/its-easy-to-judge-but-some-people-really-cant-wear-a-mask-143258

Segall B (2020) Some opponents of face mask rules turn to products offering no protection against virus. Available at: https://www.wthr.com/article/news/investigations/13-investigates/13-investigates-anti-mask-protestors-turn-to-mesh-yarn-crochet-masks-covid-coronavirus/531-5350260c-d6b1-4bd8-857e-860fe84e0f52

Sennett R (2008) *The Craftsman*. New Haven: Yale University Press.

Sheth S (2020) Trump says he thinks some Americans are wearing masks to show they disapprove of him and not as a preventive measure during the pandemic. *Business Insider*. Available at: https://www.businessinsider.com.au/trump-americans-wearing-masks-show-disapproval-not-as-preventive-measure-2020-6

Shi H, Han X, Jiang N, et al. (2020) Radiological findings from 81 patients with COVID-19 pneumonia in Wuhan, China: a descriptive study. *The Lancet Infectious Diseases* 20(4): 425–434.

Shildrick M (1997) *Leaky Bodies and Boundaries: Feminism, Postmodernism and (Bio)ethics.* London: Routledge.

Silverstone R and Haddon L (1996) Design and the domestication of ICTs: technical change and everyday life. In: Silverstone R and Mansell R (eds) *Communication by Design: The Politics of Information and Communication Technologies.* Oxford: Oxford University Press, pp. 44–74.

Silverstone R, Hirsch E and Morley D (2005) Information and communication technologies and the moral economy of the household. In: Silverstone R and Hirsch E (eds) *Consuming Technologies: Media and Information in Domestic Space.* New York: Routledge, pp. 9–17.

Singh-Kurtz S (2020) Time to master your smize. *The Cut.* Available at: https://www.thecut.com/2020/09/nailing-the-tyra-banks-smize-is-crucial-in-the-covid-19-era.html

Skof L and Holmes EA (2013) *Breathing with Luce Irigaray.* London: Bloomsbury.

Smith D (2020) 'Masks matter': Joe Biden reminds Trump that face covering is there to protect others. *The Guardian.* Available at: https://www.theguardian.com/us-news/2020/oct/06/joe-biden-masks-matter-trump-florida-town-hall

Smith J and Cha S (2020) South Korean capital orders masks on in coronavirus battle. *Reuters.* Available at: https://www.reuters.com/article/us-health-coronavirus-southkorea-idUSKBN25K04K

Southerton C and Bruce M (2019) Beyond human (un) belonging: intimacies and the impersonal in Black Mirror. *Social Beings, Future Belongings.* London: Routledge, pp. 107–121.

Sparks H (2020) These face masks come with a straw hole for sipping cocktails. *New York Post.* Available at: https://nypost.com/2020/05/12/these-face-masks-come-with-straw-hole-for-sipping-cocktails/

Staff and agencies in West Seneca (2020) Man dies after being shoved to the ground in New York mask altercation. *The Guardian.* Available at: https://www.theguardian.com/world/2020/oct/06/new-york-mask-death-man-shoved-ground

Stewart E (2020) Anti-maskers explain themselves. *Vox.* Available at: https://www.vox.com/the-goods/2020/8/7/21357400/anti-mask-protest-rallies-donald-trump-covid-19

Stokes G (2020) No shortage of surgical masks at the beach. Available at: https://oceansasia.org/beach-mask-coronavirus/

Stop AAPI Hate (2020). Available at: https://stopaapihate.org/

Stop AAPI Hate Campaign (2020) They blamed me because I am Asian. Available at: https://stopaapihate.org/wp-content/uploads/2020/09/Stop-AAPI-Hate-Youth-Campaign-Report-9–17.pdf

Surakka V and Hietanen JK (1998) Facial and emotional reactions to Duchenne and non-Duchenne smiles. *International Journal of Psychophysiology* 29(1): 23–33.

Sustainability Victoria (2020) Eco-friendly alternatives to disposable face masks. Available at: https://www.sustainability.vic.gov.au/alternatives-to-disposable-masks

TallBear K (2014) Standing with and speaking as faith: a feminist-indigenous approach to inquiry. *Journal of Research Practice*, 10. Available at: http://jrp.icaap.org/index.php/jrp/article/view/405/371

The Big Community Sew (2020) *The Big Community Sew*. Available at: https://www.bigcommu nitysew.co.uk/

The Survivors Trust (2020) COVID-19: are you concerned about wearing a mask? Available at: https://www.thesurvivorstrust.org/covid-19-are-you-concerned-about-wearing-a-mask

Thorpe H, Brice J and Clark M (2021) Physical activity and bodily boundaries in time of pandemic. In: Lupton D and Willis K (eds) *The COVID-19 Crisis: Social Perspectives*. London: Routledge, pp. 39–52.

Todd Z (2016) An Indigenous feminist's take on the ontological turn: 'ontology' is just another word for colonialism. *Journal of Historical Sociology* 29(1): 4–22.

Tronto JC (1993) *Moral Boundaries: A Political Argument for an Ethic of Care*. New York: Routledge.

Tronto JC (2013) *Caring Democracy: Markets, Equality, and Justice*. New York: New York University Press.

UCL Plastic Waste Innovation Hub (2020) The environmental dangers of employing single-use face masks as part of a COVID-19 exit strategy. Available at: https://ucl.scienceopen. com/document_file/4fbb3fa0–221b-4324–8dbf-d64efdbd26d0/ScienceOpenPreprint/ Final_Face-Mask-Paper-resumbission-29Oct20.pdf

Vallee M (2020) This is my voice in a mask. *Space and Culture* 23(3): 265–268.

Verma S, Dhanak M and Frankenfield J (2020) Visualizing droplet dispersal for face shields and masks with exhalation valves. *Physics of Fluids* 32(9). Available at: https://aip.scita tion.org/doi/10.1063/5.0022968

Vrajlal A (2020) Where to buy face masks in Australia that support Indigenous artists and migrant communities. *Huffington Post*. Available at: https://www.huffingtonpost.com.au/ entry/where-to-buy-face-masks-australia_au_5f1a27d2c5b6296fbf400258

Watson A and Bennett A (2020) The felt value of reading zines. *American Journal of Cultural Sociology* online first.

Watson A, Lupton D and Michael M (2020) Enacting intimacy and sociality at a distance in the COVID-19 crisis: the sociomaterialities of home-based communication technologies. *Media International Australia* online first.

Watts V (2013) Indigenous place-thought and agency amongst humans and non humans (First Woman and Sky Woman go on a European world tour!). *Decolonization: Indigeneity, Education & Society* 2(1): 20–34.

Wei H (2020) A brief history of face masks in China. *Sixth Tone*. Available at: http://www.sixth tone.com/news/1005177/a-brief-history-of-face-masks-in-china

Weiss-Meyer A (2020) The visible exhaustion of doctors and nurses fighting the coronavirus. *The Atlantic*. Available at: https://www.theatlantic.com/health/archive/2020/03/coronavi rus-italy-photos-doctors-and-nurses/608671/

Whittaker J (2020) Mask-making drive a success as Morwell Neighbourhood House volunteers take up cause. *ABC Gippsland*. Available at: https://www.abc.net.au/news/2020-07-31/ morvwell-volunteers-make-sure-most-vulnerable-have-face-mask/12504998

Williams MC (2020) The 15 bestselling reusable face masks on Etsy that customers give 5 stars. *Popsugar*. Available at: https://www.popsugar.com.au/smart-living/most-popular- reusable-face-masks-on-etsy-47658075

Wong E and Mozur P (2020) China's 'donation diplomacy' raises tensions With U.S. *The New York Times*. Available at: https://www.nytimes.com/2020/04/14/us/politics/coronavirus- china-trump-donation.html

Woodyatt A (2020) Museum reimagines masterpieces in the pandemic, complete with face masks. *CNN Style*. Available at: https://edition.cnn.com/style/article/art-pandemic-mask-fitzwilliam-intl-scli-gbr/index.html

World Economic Forum (2020) Do you wear a plastic mask? Here are 5 things you should know. Available at: https://www.weforum.org/agenda/2020/08/disposable-masks-plastic-pollution-coronavirus-covid-19/

World Health Organization (2020a) Timeline: WHO's COVID-19 response. Available at: https://www.who.int/emergencies/diseases/novel-coronavirus-2019/interactive-timeline?gclid=CjwKCAiA17P9BRB2EiwAMvwNyGWSa7LCiCAgb9r1TlgGmjmcYnZzOj7_zVA80ZeeVZyUsfqM35BvrhoCofQQAvD_BwE#event-7

World Health Organization (2020b) Shortage of personal protective equipment endangering health workers worldwide. Available at: https://www.who.int/news-room/detail/03-03-2020-shortage-of-personal-protective-equipment-endangering-health-workers-worldwide

World Health Organization (2020c) Coronavirus disease (COVID-19) advice for the public: when and how to use masks. Available at: https://www.who.int/emergencies/diseases/novel-coronavirus-2019/advice-for-public/when-and-how-to-use-masks

World Health Organization (2020d) WHO Director-General's opening remarks at the media briefing on 2019 novel coronavirus, 7 February 2020. Available at: https://www.who.int/dg/speeches/detail/who-director-general-s-opening-remarks-at-the-media-briefing-on-2019-novel-coronavirus-7-february-2020

World Health Organization (2020e) Advice on the use of masks in the context of COVID-19: interim guidance, 6 April 2020. Available at: https://apps.who.int/iris/handle/10665/331693

World Health Organization (2020f) Advice on the use of masks in the context of COVID-19: interim guidance, 5 June 2020. Available at: https://apps.who.int/iris/handle/10665/332293

World Health Organization (2020g) When and how to use masks. Available at: https://www.who.int/emergencies/diseases/novel-coronavirus-2019/advice-for-public/when-and-how-to-use-masks

World Health Organization (2020h) Coronavirus. Available at: https://www.who.int/health-topics/coronavirus#tab=tab_3

World Health Organization (2020i) COVID-19 mythbusters. Available at: https://www.who.int/emergencies/diseases/novel-coronavirus-2019/advice-for-public/myth-busters

Zangmeister CD, Radney JG, Vicenzi EP, et al. (2020) Filtration efficiencies of nanoscale aerosol by cloth mask materials used to slow the spread of SARS-CoV-2. *ACS Nano* 14(7): 9188–9200.

Zhou M, Yu Y and Fang A (2020) Asians in US torn between safety and stigma over face masks. *Nikkei Asia*. Available at: https://asia.nikkei.com/Spotlight/Coronavirus/Asians-in-US-torn-between-safety-and-stigma-over-face-masks

Zhou YR (2020) The global effort to tackle the coronavirus face mask shortage. *The Conversation*. Available at: http://theconversation.com/the-global-effort-to-tackle-the-coronavirus-face-mask-shortage-133656

Index

https://doi.org/10.1515/9783110723717-010

www.ingramcontent.com/pod-product-compliance
Lightning Source LLC
Chambersburg PA
CBHW052138270326
41930CB00012B/2941